By the same author

Managing Canal Irrigation in India : Problems and Their Resolutions

---

About the Author

Sib Ranjan Misra, a Ph.D. in Economics (Jadavpur University) and D.E.P. (Istituto di Studi per lo sviluppo Economico, Naples University, Italy) and a Japan Foundation Fellow (2004), was on the teaching staff of the University of North Bengal (1971-78), currently Professor of Economics, Department of Economics and Politics, Visva-Bharati, Santiniketan. His present assignment includes research and teaching in various branches of Economics. He has completed quite a number of projects on problems of tea plantations, fisheries development, irrigation management etc. sponsored by ICSSR, UGC, ICAR etc. He has also served for quite some time as an Adviser on the Fisheries Development Advisory Board, Ministry of Agriculture and Co-operation, Government of India. The author's publications have appeared in various professional journals and also include books entitled: *Tea Industry in India* (1986), *Fisheries in India* (1989), *Floriculture in West Bengal—Problems and Potentials* (co-authored, 1997) and *Managing Canal Irrigation In India* (2001).

INLAND FISHERIES IN INDIA

INLAND FISHERIES IN INDIA

Issues and Concerns

Sib Ranjan Misra

CONCEPT PUBLISHING COMPANY, NEW DELHI-110059

Cataloging in Publication Data—DK
Courtesy: D.K. Agencies (P) Ltd. <docinfo@dkagencies.com>

Misra, Sib Ranjan, 1947-
Inland fisheries in India : issues and concerns / Sib Ranjan Misra.
p. cm.
Includes bibliographical references (p.)
Includes index.
ISBN 8180693104

1. Fisheries—India—West Bengal. 2. Fisheries—Government policy—India—West Bengal. I. Title.

DDC 639.209 541 4 22

ISBN 81–8069–310-4

First Published 2006

Published and Printed by
Ashok Kumar Mittal
Concept Publishing Company
A/15-16, Commercial Block, Mohan Garden
New Delhi (INDIA)

Phones : 25351460, 25351794
Fax : 091–11–25357103
Email : publishing@conceptpub.com

Dedicated
to
My Parents

Go to the people
Live with them
Learn from them
Love them

Start with what they know
Build with what they have

But with the best leaders
When the work is done,
The task accomplished.
The people will say,

"We have done this ourselves"

—Lao Tsu (China),
700 B.C.

Preface

There has been a growing recognition of the vast growth potential of production of fish, mainly from inland water bodies, thereby providing the much-needed edible protection and augmenting the income of the rural poor. India has abundant natural endowments, conventional wisdom and human skills in respect of intensified fish farming. But in spite of enormous potentials, much headway has not been made. The levels of production and productivity have not been adequate, leaving a large gap between actual and potential yield. The new technology, which seems technically attractive, has failed to produce substantial results in real world. The recipients of technology are greatly circumscribed by the factors like multiple ownership of tanks and ponds, shortage of quality fish seeds, shortage of investment capital, inadequate extension work, poaching, deliberate poisoning inadequately organized marketing facilities and many more others. In other words, the present unsatisfactory performance may be attributed mainly to absence of rational management and judicious utilization of the resources.

In this study, attempts have been made to examine the operational features of tanks and ponds and to identify the factors that impede the development of fisheries in West Bengal. Efforts are also made to evaluate the role of government programmes towards development of inland fisheries sector. The study is based on the data collected from selected fish farmers from different villages in the selected agro-climatic zones of West Bengal. This monograph is largely based on the Report of a Major Research Project on the Economics of Pisciculture which was undertaken with the financial support from the University Grants Commission. I express my deep sense of gratitude to the UGC for their support.

I wish to express my thanks in abundance to Shri Jaydev Misra, a Ph.D. student in our Department who consistently took althrough the trouble of collecting and collating the data and preparing the statistical tables appropriate for this exercise. Shri Misra also carefully wordprocessed the manuscript and saw to it that I met the deadline.

Through my frequent interactions with the Professors like Kyo Kambayashi, Tokyo Metropolitan University, Professor Ichiro Horide, Professor Yukimasa Nagayasu, International School of Economics, Reitaku University, Dr. Kuniyo Tsubota, Director, Asian Productivity Council, Tokyo; Professor Hiroike Masanaro, Hitotsubasi University during my stay in Japan as a Japan Foundation Fellow. I have been able to enhance my understanding about the different dimensions of Indian aquaculture. I owe a special debt of gratitude to all of them for shaping my thought about it.

A word of warm appreciation is due to Sabhadipatis of Murshidabad, Malda, Bankura and South 24-Parganas, Agriculture Development Officer and members of the Panchayat Samitis of the Study area. Our obligation is due to Director, Directorate of Fisheries, West Bengal; Dr. Goutam Sarkar, Deputy Director of Fisheries (Central Zone); Dr. Samik Das, Chief Executive Officer, FFDA, Murshidabad; Sri A.K. Gorai, Fishery Extension Officer, Malda and Sri N. Barik, Assistant Director of Fisheries, Birbhum who provided us with most of the facts and figures and many of the ideas which are set out below. It is thanks to their expertise and unfailing courtesy. To the numerous fish farmer respondents of our study, we owe a debt which no expression of gratitude can fully redeem. We like to acknowledge their contribution in making the study what it is.

The staff of the Libraries (Central Library, Visva Bharati and Agro-Economic Research Centre, Visva Bharati), where we spent most of the time involved in completing this work gave us their full co-operation. We are thankful to them, particularly to Dr. Kaji M.B. Rahim, Principal, Palli Siksha Bhavan; Prof. Kalyan Brata Bhattacharya, Dr. Kaushik Gupta, Prof. Soumyen Dutta in the Department of Economics, Burdwan University and Dr. Madan Gopal Ghosh, Ex Director and Sri Debangshu Majumdar, Agro Economic Research Centre, Visva Bharati. We are thankful to the staff members of the Department of Economics and Politics in Visva Bharati for their continuous academic support during the course of this investigation and finalization of this report.

I am really grateful to Shri Ashok Mittal of the Concept Publishing Company who kept me always going and for his indefatigable interest in bringing out this monograph to the market. In the process, I found a very good friend in him.

Sib Ranjan Misra

Contents

List of Exhibits and Figures

Exhibits

Figures

1

Introduction

Give a man a fish and he eats for one day; teach him how to fish and he eats for a lifetime. —Henry David Thoreau

India has the distinction of being one of the seven largest producers of fish in the world and ranks second in terms of total inland fish production after China. All in all, this sector has been recognized as a powerful income and employment generator since it stimulates the growth of a number of allied industries and also is a source of cheap animal protein. It is an instrument of livelihood for a large section of economically backward population of the country.

India has an estimated overall fish production potential of about 9 million tonnes, with 4 million tonnes expected from marine fisheries. As against the said potential the country is producing around 6 million tonnes (2001-02) with a contribution of about 50 per cent each from marine and inland fisheries. Based on the total production, India occupies third position in the world contributing a little over 4 per cent of the world's total production. In the case of inland fish production India ranks second in the world. The current per capita availability of fish per annum works out to about 9 kg for the fish eating population of the country which is less than their minimum requirement of 11 kg recommended by the World Health Organization. The structure of fish production in the country has undergone significant changes over the years. In 1950-51, the share of marine fisheries was 71 per cent of the total fish production of 7.52 tonnes. During the 1970s and 1980s, the share of inland fisheries was just above one-third of the total fish production. Later, it gradually increased to reach nearly 40 per cent in 1980-81 and more than 50 per cent in 2000-01. It is sometimes argued that the increase in the share of inland fisheries was, on the one hand, due to deceleration in growth in marine fish production and on the other because of a policy shift in favour of inland fisheries, especially, aquaculture (Ayyappan, 2004).

Thirty-five per cent of population in India has been estimated to be fish eaters; being 46 per cent in the coastal and 27 per cent in the non-coastal states. Wide regional variations in fish consumption are being observed. There are some studies which mentioned that the responsiveness of demand for fish to changes in price and income/ expenditure differs for fish across income classes, regions and urbanization. With increase in population and per capita income, the demand for fish will increase in the coming years. Own price elasticity of demand for fish is high and has not demonstrated any sign of decline during the last 15 years. These elasticities are substantially higher in the eastern and northeastern states. This calls for concerted technological innovations and institutional re-organizations in the fish sectors to increase the overall welfare of both the producers and the consumers.

. Indian fisheries constitute an important sector of our national economy for various reasons. The importance of this sector can be better appreciated in the light of the fact that this sector provides wholesome protective food containing animal protein in the dietary of the country's undernourished population. In the country like India, where half of the population still does not get two square meals per day, it is important to pay attention to any alternative source which, besides being within the reach of the people, will constitute a nutritive diet.

Inland sector remains a sector of much promise. Unlike agriculture, because of expansion of culture based fisheries enterprises, the contribution of inland fisheries sector to grow domestic product has been increasing at a significant rate. This sector is complex enterprises which operates under the intensified network of natural resources, other enterprises with fisheries and other socio-political variables. This is evident from the following exhibit that during the period1995-96 to 2001-03 the percentage of possible changes in output values with respect to time has mined by 30 per cent with inland fisheries, constituting more than 55 per cent of the growth. The rate of growth of value of output of the fisheries sector has been targeted 7 per cent per annum during the Fourth Five Year

Plan (GOI, 1999). Fish consumption is also expected to increase from 5.05 kg/capita in 1996-97 to 9.30 kg/capita in 2011-12 (GOI, 1999).

Exhibit 1: Change in Value of Output in Fisheries, Agriculture etc.

Sector	*1995-96*	*1996-97*	*1997-98*	*1998-99*	*1999-2000*	*2000-01*	*2001-02*
1	*2*	*3*	*4*	*5*	*6*	*7*	*8*
Agriculture	103.60	113.65	110.19	118.68	119.43	116.43	122.40
Fisheries	112.25	117.77	120.75	128.20	139.70	140.78	145.78

Source: National Account Statistics, Central Statistical Organization.
Note: This is of course worked out on the basis of expected population growth rate in per capita consumption expenditure and the elasticities of demand with respect to per capita consumption expenditure.
Reference: Government of India (1999), Ninth Five Year Plan, 1997-2002, Vol. I. Development Goals—Strategy and Policies, Planning Commission, New Delhi.

According to National Income Statistics, Indian fisheries contribute about 1.4 per cent of the total Gross Domestic Product (GDP) and 4.73 per cent of the GDP from agriculture.[1] The vast resources which are capable of yielding 10 million tonnes of fish per year as against only an average annual production of 5.65 million tonnes per year will go a long way in increasing the national income.

Foreign exchange earnings from export of fish and fishery products have been steadily rising during recent years. In absolute term, the export earnings have steadily increased from Rs. 234.84 crores in 1980-81 to Rs. 5,114 crores in 1999-2000 which accounted for 1.76 per cent of the total export of the country.[2] It is also estimated that with adequate facilities, the export earnings can be more than doubled within three or four years.

It is generally difficult to assess correctly the population engaged in Indian fishing industry because of the general inadequacy of statistical data on fishermen population, the system of casual labour prevalent in the industry, the difficulty in distinguishing fishermen wholly or partly dependent and fishermen directly or indirectly dependent on this industry. However, an important advantage of employment opportunities in this sector is that it is mostly the rural section of the population which is benefited most.[3] It is also observed from the Census of India data that about 73 per cent of the Indian fishermen and related workers are engaged in the rural sector. Heavy pressure of the population on arable land, coupled with conditions of the backward technology and organization has resulted in severe unemployment and low output per engaged person. Pisciculture may provide some additional employment. Most importantly, it is the source of livelihood for a large section of economically backward population of the country.

Public revenue from fisheries would be welcome in the context of present financial position of a large number of Indian states. Leaving aside the justification or otherwise of making fisheries as a source of public revenue, it is important to evaluate the revenue potentialities of fisheries. The revenue from fish culture has reached a level of about 20 per cent in the case of Gram Panchayat of some states like Orissa, Uttar Pradesh.

The fishing industry, when developed on a commercial scale, gives rise to a larger number of ancillary industries. Important among them are shipbuilding, marine engines, nylon manufacturing, marketing, freezing, canning, fishmeal and fish oil, cold storage, ice manufacturing and various others. It is difficult to formulate an idea about Indian fishing ancillary industries through the published data. But it can be stated that a state like Kerala alone has 24 cold storages, 21 freezing factories, 40 boat building yards etc.[4] It should be noted that USSR has invested more than half of her food industry investments since 1952 into the fishing industry.[5] The potential forward and backward linkages through these activities contribute to diversification and strengthening of the regional and national economy.

In sum, the development of fishing and allied industries in India has a special significance, because –

- this sector is eminently suited to assist a large mass of backward and economically weaker section of rural community,
- it provides a huge protein subsidiary food at reasonable prices to satisfy certain rigid consumer preferences, and
- it has a large export potential.

The fisheries of India broadly consist of two types – (a) marine fisheries and (b) inland fisheries that include fresh water, estuarine and backwater fisheries. Of the total fish production of 5.66 million tonnes during 2000-01, about 2.84 million tonnes came from the marine sources and the rest from the inland water. Inland fisheries resources consist of two types of water – brackish and fresh. The fisheries of this type of water can be further classified into two types—culture and capture fisheries. Capture fisheries consist of rivers and their tributaries and estuaries, lakes etc., wherein fisheries are harvested without having to sow the seeds, nature is doing the job of sowing. In case of culture fisheries, one has to sow the seeds, nurse it and tend it to a proper size before harvesting. Culture fisheries comprise ponds, tanks, swarms, marshes etc. These different segments—capture, culture and culture-cum-capture fisheries have different settings and require entirely different types of inputs, infrastructure and development strategy.

India has a vast potential for development of inland fisheries. Indian fishery resources are vast and varied comprising rivers, canals (1,91,000 km), reservoirs (3.15 million hectares), ponds and tanks (2.25 million hectares), floodplain, lakes and derelict waters (1.3 million hectares) and brackish water areas (1.42 million hectares). Marine fisheries resources in the Indian EEZ (Exclusive Economic Zone) have been estimated at about 3.934 million tonnes, constituting 51 per cent demersal, 43 per cent pelagic and 6 per cent oceanic.

The performance of the fisheries sector is quite consistent. It is quite evident from Exhibit 1.1 that fish production was 0.75 million tonnes in 1949-50. It increased to 2.34 million tonnes during 1979-80 and then to 3.68 million and 5.26 million tonnes in 1989-90 and 1998-99 respectively. In terms of growth, production rose by 4.78 per cent during the period from1979-80 to 1989-90 and then by 4.70 per cent during the period from 1989-90 to 1998-99.

Again, the structure of fish production in the country has undergone significant changes over the years. During the 1970s and 1980s, the share of inland fisheries was just above one-third of the total fish production. Later, it gradually increased to reach nearly 40 per cent in 1990-91 and more than 50 per cent in 2000-01 (Exhibit 1.2)[6]. Fish consumption is likely to increase from 5.05 kg per capita in 1996-97 to 9.30 kg per capita in 2011-12 (Government of India,

1999). By 2020 AD, the domestic demand for fish is expected to be anywhere between 125 lakh tonnes and 200 lakh tonnes.[7] The inland

Exhibit 1.1: Production and Growth Rates of Some Major Agricultural Commodities

Commodity	*Production (million tonnes)*				*Growth Rates (per cent per annum)*					
	1949-50	*1979-80*	*1989-90*	*1998-99*	*1979 to 1989-90*			*1989-90 to 1998-99*		
					Area	*Production*	*Yield*	*Area*	*Production*	*Yield*
1	*2*	*3*	*4*	*5*	*6*	*7*	*8*	*9*	*10*	*11*
Rice	23.54	42.33	73.57	85.99	0.45	4.29	3.82	0.49	1.70	1.21
Wheat	6.39	31.83	49.85	70.78	0.57	4.24	3.65	1.64	3.64	1.96
Food grains	54.92	109.70	171.04	203.04	-0.11	3.54	3.33	-0.18	1.80	1.42
Fish	0.75	2.34	3.68	5.26	—	4.78	—	—	4.70	—

Source: Tenth Five Year Plan, 2002-2007, Vols. II, III, Planning Commission, Government of India, New Delhi.

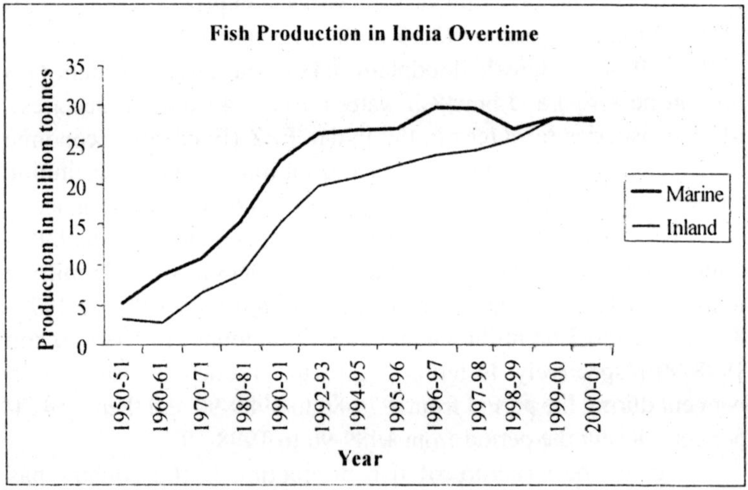

Fig. 1.1: Fish Production in India

Source: ME & MS Wing, Directorate of Fisheries, Government of West Bengal.

fisheries production will have to increase sevenfold even if minimum expected demand is to be satisfied from the inland sector. In this situation, it is essential to fill the supply-potential gap rather than supply-demand gap.

With a catchment area of over 3.12 million km, 14 large, 48 medium and innumerable small rivers carrying a surface run-off of 167.23 million hectares through their drift exhibit a kaleidoscopic variety of environments, such as the cold-streams of the uplands, fast and slow moving stretches in the plains, floodplain lakes, deep pools, placid backwaters and estuaries (Exhibit 1.2). Development and rational exploitation of our capture fisheries resources are challenging tasks, unlike culture fisheries, where ecosystem manipulation and production monitoring are easy. Large open waters, spread among varied agro-climatic conditions exhibit diverse geo-dynamics and hence no uniform guideline can be formulated for their managements.

Exhibit 1.2: Inland Water Resources of India

No.	*Resources*	*Quantity*
1.	Rivers (km)	29,000
2.	Irrigated Canals (km)	1,20,000
3.	Flood Plains Lakes (ha)	2,02,213
4.	Upland Lakes (ha)	72,000
5.	Reservoirs (ha)	3,15,798
6.	Estuaries, Mangroves, Brackish Water Lagoon	60,00,000
7.	Estuaries Wetlands (ha)	50,000
8.	Brackish Water Aquaculture	9,02,000
9.	Fresh Water Aquaculture	7,53,000

Source: Handbook of Fisheries Statistics, Government of India.

West Bengal is well endowed for inland water fisheries. It is situated in the monsoon belt. It has an annual precipitation of about 1000 mm. It has extensive inland water resources in the form of hill streams, rivers, canals, wetlands, ponds and tanks[8] etc. According to the West Bengal State Statistics (Directorate of Fisheries), the following (Exhibit 1.3) are the water resources available in the State.

West Bengal produces the largest quantity of fresh water fish. About 65 per cent of total production of fish comes from fresh water, mainly, tanks and ponds. The estimated demand for the State is eight lakh tonnes. The shortfall is made up to an extent by import of fish from other states.

The Government of West Bengal has recently taken steps for intensifying fish culture and a considerable investment has already been committed. Fish Farmers' Development Agencies (FFDA) which have been established in the State have extended facilities like fish seed supply at cheaper rates and subsidy. Extension services have been geared at village level. Training programme for the fish farmers has been introduced. Despite all these efforts and enormous potentials for intensified fish farming, these water bodies have not been utilized optimally for fish culture. The production level and productivity are still not adequate to meet the large gap between potential and actual yields. Improved technology has not yet disseminated to the level of

Exhibit: 1.3: Water Resources in West Bengal

Tank/Pond	2,76,201.90 ha
Beel/Baor	41,781.65 ha
Rivers	1,72,586.36 ha
Reservoirs	16,738.80 ha
Sewage Fed	4,083.00 ha
Canal/Creek	30,000.00 ha

Source: ME & MS Wing, Government of West Bengal.

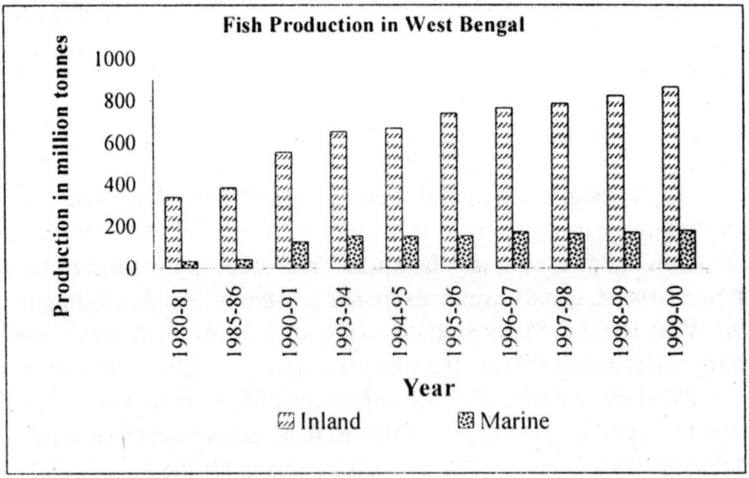

Source: ME&MS Wing, Government of West Bengal.

Fig. 1.2: Fish Producion in West Bengal

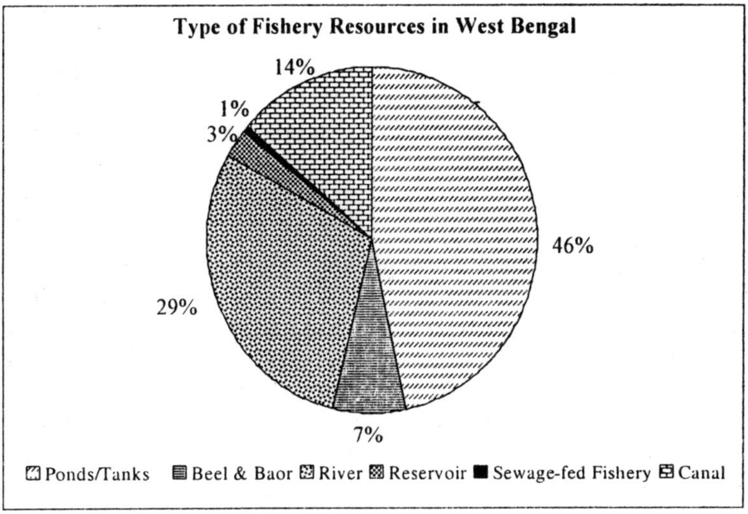

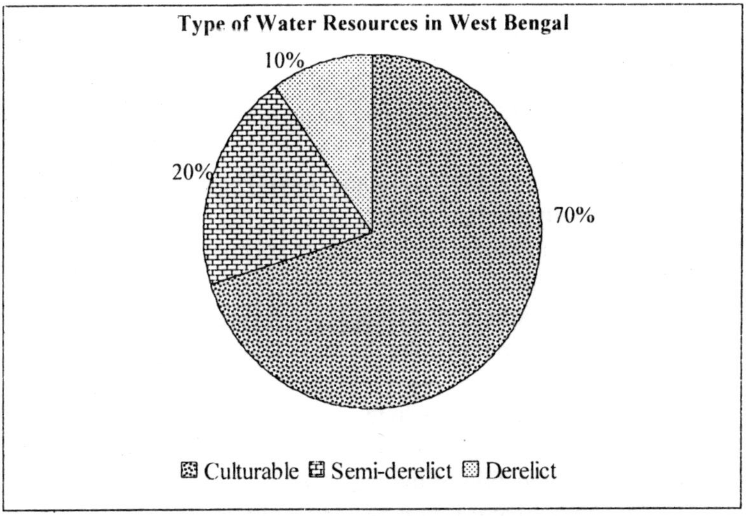

Fig. 1.3: Different Types of Water Resources in West Bengal

common fish farmers in the villages. The recipients of technology are generally circumscribed by different factors. When the technology has been developed and resources available, one should expect reasonably that production should increase several folds in the next few years. There is an urgent need to make institutional efforts in the field of research, extension and training to exploit the fish production potentiality. This, in fact, has been in utter negligence in our country. There is hardly any comprehensive research in the field of fish culture in general and economics of inland fish production in particular.

Objectives

The objectives of the present study are:

- To examine the operational features of tanks and ponds,
- To decipher the factors that impede the development of the pisciculture in fresh water ponds and tanks,
- To evaluate the performance of the existing government policies, and
- To suggest some remedial measures.

Hypotheses

The corresponding Hypotheses may be as follows:

I. The water-bodies (tanks and ponds) have not been utilized optimally for fish culture.
II. The technologies have not been adequately adopted by the fish farmers.
III. The technological changes and government programmes have not generated the desired yield.

The process variables include:

- Size and location of the water units.
- Ownership pattern of tanks and ponds.
- Availability of investment capital.
- Poaching, deliberate poisoning and some other social problems.
- Inadequately organized marketing channels.
- Identification of beneficiaries of Government programmes.
- Subsidy for tanks improvement and inputs like seed, feed etc.

The impact variables (an outcome of the process variables) are:

- Actual water area brought under cultivation.
- Yield per acre.
- Production, cost and income from aquaculture.
- Price realization.
- Price spread.
- Employment generated etc.

With the help of the above mentioned variables, efforts will be made to articulate the proposed objectives.

Scheme

The Scheme of the study is as follows. In chapter **One**, the importance of this sector in our national economy has been sketched. Efforts are also made to explain, in short, the Indian fisheries resources. Explanation about the scope and objectives are also made in this chapter. The **Second** chapter deals with a short review of the literature. Attempts are also made in chapter **Three** to explain the methodology of the study. In chapter **Four**, structure and operational features of the tanks and ponds are explained. Chapter **Five** offers a systematic overview of the adoption behaviour of the different packages of practices. In this context, some of the major constraints faced by the fish cultivators are also explained. Chapter **Six** provides an explanation of the role and inadequacies of the government programmes, precisely like the Fish Farmers Development Agencies (FFDAs) in the development of inland fish cultivation. In the final chapter, **Seven,** the main conclusions of the study are summarized and the policy implications thrown up by the study are also elucidated.

NOTES

1. National Income Statistics, Central Statistical Organization, Government of India, 2004.
2. Export Prospects of Fish and Fish Products: National Council of Applied Economic Research, New Delhi.
3. *Fishing News International,* July 2004, pp. 21-22.

4. Kerala Fisheries—Facts and Figures: 1999-2000, Directorate of Fisheries, Government of Kerala, Tiruvananthapuram.
5. *Fishing News International,* Vol. 5, No. 10, pp. 10.
6. It is sometimes argued that the increase in the share of inland fisheries in India was, on the one hand, due to deceleration in the growth of marine fish production and on the other, because of a policy shift in favour of inland fisheries, especially aquaculture. Again, the strong traditional knowledge base supported by scientific inputs in various aspects of management of inland fish cultivation is one of the factors responsible for growth in inland fish production (Jana and Sena, 2004).
7. 35 per cent of population in India has been estimated to be fish eaters, it being 46 per cent in the coastal and 27 per cent in the non-coastal states. With increase in population and per capita income, the demand for fish in India is likely to increase in the coming years. Own price elasticity of demand for fish is high and has not shown any sign of decline. There are some studies which mentioned that the responsiveness of demand for fish to changes in price and income (expenditure) varies for fish across income classes, regions and urbanization (Ayyappan, 2004).
8. The preponderance of small tanks and ponds and the existence of large number of derelict and semi-derelict ponds in West Bengal are quite evident from Annexure 4.

2

Review of Literature

Economic information on the various aspects of inland fish cultivation have seldom appeared in literature, though much research works have been done in the biological and marine sciences. It is not properly co-ordinated with the economic ideas. One of the major reasons for the unsatisfactory performance may be low availability of appropriate reliable data and standardized terminologies necessary for comparability and aggregation of data over time and across regions (Srivastava and Reddy, 1983).

B.S. Saxena (1968) observed that lack of detailed data on resources, their utilization, production levels and marketing practices have affected adversely the development of the inland fish cultivation. He suggested that the steps should be taken to improve credit arrangements to the farmers.

V. Gautam (1983) pointed out that the inland fisheries are equally vast and diverse and potentially one of the richest in the world. The progress and development of the inland fisheries have so far been impressive. There is greater potential in still water resources than in flowing water. His study highlights the points that fish production in ponds and tanks is much higher than the larger water reservoirs and lakes resources. He estimated that by 2000, the domestic demand for fish is going to be 200 lakh tonnes. Even if expected demand is to be satisfied from the inland sector, inland fisheries production will have to increase seven times in next two decades.

He suggested that there is a need to practise aquaculture in combination with agriculture. But unfortunately, he did not take care of the financial condition of the poor fishermen, inadequate supply of better inputs and at reasonable rate.

Dineshwar Prasad (1968) in his "Economic Aspects of Fisheries Development" made an attempt to investigate that the economic problems of fisheries development normally arise from various factors, e.g. a high degree of risk and uncertainty involved in fish production. Over and indiscriminate fishing of juvenile fishes is another problem. As the fishermen sometimes may not be the owner of the ponds and tanks, the fishermen harvest and sell their product to minimize their risk. Generally, the fishermen do not have any long-term interest in stocking of large and matured fish. They have a tendency to maximize yield at the earliest shortest possible time. Among numerous other problems, Prasad pointed out mainly the factors like existence of middlemen, lack of governmental incentives, sale of undersized fishes, general poverty and low purchasing power of the people and absence of organized centres for fish marketing. Measures suggested by him are proper exploitation of huge water resources through effective planning like appropriate stocking, proper manuring of tanks and training of the fishermen for improved method of fish rearing, catching and storage were also suggested.

S.K. Chakrabarty (1968) observed that inland fisheries, particularly pond culture, would provide the best opportunity for rapid increase in fish production—because fresh water in tropical regions has a higher rate of production per unit area than the brackish water. According to his estimate, production in the study area (a village in Birbhum district of West Bengal) was fairly very small. The main reason behind low productivity of fish production identified by him are deplorable condition of the tanks and ponds as also their low level of maintenance as the large number of tanks and ponds were weed infested and semi-derelict. Due to continuous deposition of silt in the bottom layer of the water area – the average water level has been gradually going down. He suggested that there is a need for immediate desiltation of the tanks and eradication of excess vegetation and water hyacinth to maintain the mere existence of as many as two-thirds of the tanks. Sometimes, it was found impossible to carry on fish cultivation as water dries up completely in the summer.

Measures suggested by him for improvement of production are the formation of Fisheries Co-operative Societies. Institutions like Directorate of Fisheries and Local Administration (Panchayat) can play an effective role in formation of such societies. He emphasized that demonstration of successful fish farms in the matter of the adoption of improved techniques will provide the incentive to other fish farmers to go in for the adoption of scientific packages and hence, higher yields. Among other measures, he also suggested measures like provision of adequate financial help in the forms of grants and loans, effective legislation for better and efficient management of water bodies. In case of multi-owned ponds, by legislation, an efficient management should be established in order to obtain and maintain a minimum level of production. Another important measure suggested was that for higher production, tanks should be leased out to a single person on the guarantee that he will produce the minimum level of production and for a reasonable period of time.

In the book *Inland Fisheries Resources in India* (1983) by **Srivastava** and **Vasistha**, published by the Concept Publishing Company, New Delhi and *Fisheries Development in India* by **Srivastava** and **Dharma Reddy**, published by the Concept Publishing Company, New Delhi, analysis on nature and problems of fisheries, both inland and marine, have been done. In the articles, like "Economic Aspacts of Fisheries Development of Bihar" by **D. Prasad** in *Indian Journal of Agricultural Economics* (*IJAE*, 1968), Vol. XVIII; "Fisheries Co-operative in India" by **Bhalara** and **Kali Charan** in *Indian Journal of Agricultural Economics (IJAE)*, Vol. XVIII; "Benefit-Cost Ratio in Fish Culture" by **Dhondy** and **Singh** in *IJAE,* Vol. XVIII, attempts were made to identify some of the basic constraints.

In the monographs like "Potentials and Problems of Composite Fish Farming in West Bengal" by **S.M. Murshed**, *CIFRI Bulletin* No. 5; "Scope for Scientific and Commercial Fish Farming in West Bengal" by **A.V. Natarajan**, *CIFRI Bulletin* No. 9 and "Export Prospect of Fish and Fish Products" (1986) by NCAER, New Delhi, some of the issues relating to the problems of marketing of fish production have been dealt with.

Some economists have tried to highlight this issue on their articles. **G.S. Gupta** (1983), in his article "Freshwater Fisheries—Present Status, Prospects and Policy Measures" showed that there is an inverse relation between pond size and yield rate. That is, as pond size increases, the productivity decreases. The conclusion was derived from the data, which is based on the new concept of Effective Storage Level (ESL). On the basis of its biological implication, it has brought two other concepts—(i) full storage level and (ii) dead storage level

(DSL). Still, it can be said that the conclusion is not always valid, as it did not consider input-output ratio. It may happen that if inputs are increased along with the pond size, rate of fish production increases by the same amount or more than the input growth, depending on whether there is increasing or constant returns to scale in inland fish production.

Opposite view regarding farm size-productivity relation in Pisci-cultures was also explained by some other researchers. At least two of them deserve particular mention. **Vatshala** (1983) opines against the assumption of negative correlation of pond size and productivity debate. According to Vatshala, in a big pond, if cultivators take intensive care, by incorporating various improved technologies and with proper training, then it is not difficult to get a positive return from big size ponds.

Ram Charan (1983) in his study makes a point that there is not much difference between composite fish culture, yield and traditional yield. He assumes that it may be important and essential for maintaining the eco-system, but as regards its economy, it certainly does not make any difference whether we have poly-culture, monoculture or composite culture.

A.J. Singh and **V.K. Gupta** (1983) dealt with the problem in relation to demand, supply and production, marketing and distribution of fish seed. In their in-depth study in West Bengal, the biggest seed producing and exporting state in India, they have given a clear picture about the infrastructural developments taking place in seed production. The authors then would not have left the notional idea that the gap between existing supply and demand in the nineties would be highest in West Bengal.

The study conducted by **I.I.M.A.** (1983) has made some recommendations for the development of pisciculture. These recommendations like database and statistics, research on specific problems of commercial fish culture issues like classification of districts into good, medium and poor on the basis of some criteria of potential and stage of development and adequate supply of quality fish seed, more intensive stocking and partial stocking and partial harvesting at intervals, long leasing system and fixation of lease amount, improvement of fish farmers co-operative for supply of inputs, harvesting and marketing, improvement of Fish Farmers

Development Agencies (F.F.D.A.) which have significant bearing upon fish cultivation have been clearly explained.

R.K.P. Singh (2003) in his book entitled *Economics of Aquaculture*, published by Daya Publishing House, New Delhi, made an attempt to study the various socio-economic aspects of inland fish farming enterprises in Bihar. The specific objectives were to examine the culture practices, costs, returns, labour, and employment, marketing practices and constraints in the fish production, on the basis of primary data. He identified some of the institutional and technological constraints like low level of input use, non-existence of fish nurseries, lack of institutional credit etc.

In an article entitled "Input Resource Management in Fisheries" published in *Indian Journal of Agricultural Economics* (IJAE), **Katiha** *et al.* (1988) attempts to explain the sub-optimal utilization of inputs. There exists scope for increasing fish production if inputs are used optimally. The marginal value productivities of inputs like gear weights, fishing efforts are sufficiently large. They employed the standard Cobb-Douglas type of production functions to estimate the elasicities of inputs.

Radheysham (2001) in an article "Community based Aquaculture in India" in Aquabite studied the relevance of the community-based aquaculture founded on the principles of common interest groups working together regardless of sex and age. This has been an effective tool for implementing scientific aquaculture programme in India. He used Strength, Weaknesses, Opportunities and Threats (SWOT) method.

U.K. Srivastava (1990), along with other economists from Indian Institute of Management, Ahmedabad in the Report "Fish Farmers' Development Agency Programme for Fresh Water Aquaculture" made a thorough study about how the large numbers of under-utilized fallow culture fishing resources could be brought under optimum fish production in the area of its operation. They have exhaustively analyzed the primary data they had collected. They arrived at the conclusion that a large number of institutional and technological factors are there to affect adversely the production of fish. The suggestions they made were to rationalize the leasing arrangements, to provide subsidy and effective training programme, to arrange smooth supply of fish seeds at reasonable prices and also to develop the marketing facilities.

Another important aspect of pisciculture is about multiple ownership of most of the private ponds and tanks. Most of the private tanks and ponds are not owned by the single owner. People, particularly in West Bengal have incentives to hold the percentage of ownership of inherited ponds with their other family members and division of the ownership continues from one generation to next generation. Thus, as generation proceeds, share of ownership by single owner decreases, which gives rise to a very complex plurality of ownership. Such problem often impedes new investment for the improvement of ponds and tanks and also the intensive application of new technologies. **M.M. Bhalerao** and **Kalicharan** (1968) recognized the problems of existence of plurality of ownership and interference of middlemen as senior deterrents.

Sib Ranjan Misra (1987), in his book *Fisheries in India,* published by Concept Publishing Company reveals that poaching acts as a major constraint leading to poor performance of fish production particularly in West Bengal tanks and ponds. S.R. Misra, on the basis of data collected from tanks and ponds' owners in three districts of West Bengal, arrived at the conclusion during 1983-84 that poaching or deliberate poisoning due to rivalry, enmity or jealousy are widespread. Among other problems, lack of proper transport facility, suitable marketing management, non-availability of opportunity for soil and water testing etc. were considered by the poor fishermen as the major inhibiting factors. In consideration to non-adoption of new techniques, he observed that lack of information regarding various improved practices is one of the important factors. The financial difficulties due to bureaucratic delay, etc. make the practice of composite fish-farming technologies much more difficult. Other problems emphasized by Misra were erratic and inadequate supply of quality fish seeds or fingerlings in time and at reasonable rate, multiple ownership and adulteration of inputs etc. According to Misra, this unsatisfactory performance can be attributed mainly to absence of rational management, judicious utilization and appropriate exploitation of the resources.

Measures suggested by Misra are the utilization of canal irrigation (from which water can be released to the fish ponds in winter season) or rain fed farming on water shed basis, desilting of about to dry-up ponds, long term lease arrangement, government legislation to evict multiple ownership, development of appropriate technologies (keeping in view the regional, environmental and ecological differences), organized marketing channels, mixed farming etc. Pumping of more money into the economy may not necessarily generate the desired result. According to Misra, more village-based co-operatives should be set up and existing ones should be reformed and galvanized.

A.V. Natarajan (1985) from a study at 24-Parganas observed that a number of tanks and ponds are used for seasonal irrigation purposes. An antagonistic relation for water availability for irrigation and for the purpose of pisciculture seriously affects fish cultivation.

An in-depth study by **Agro-Economic Research Centre** (1986), Visva-Bharati, Santiniketan in its "Use of Tanks for Minor Irrigation and Pisciculture in West Bengal" tried to explain the actual position of uses of tanks for irrigation and fisheries purposes. The main findings of the study are that the scope of using tanks for irrigation is rather limited; while potentialities of pisciculture are quite large. Some institutional and legal reforms are urgently needed to overcome the existing constraints. Improvement of the technical characteristics of tanks like depth, siltage, weeds etc. are of crucial importance for long-term improvement of tanks. The importance of existing social forestry in the embankment of tanks has also been highlighted in the study. The result was derived by examining the effects of some 'cause' (independent) variables on a number of dependent variables.

K.S. Chattopadhyay (1987) in his "Problems and Prospects of Pisciculture" discussed various institutional, technical and financial factors, which are found detrimental to smooth functioning, and augmentation of growth of fish production in a district like Birbhum in West Bengal. He was optimistic that if such vast water bodies are properly exploited, both the levels of production and productivity of fish in this district could be increased several folds. But he severely criticized the role of the government while rendering extension programmes at the grassroots level. Functioning of F.F.D.A. in his study area was also subjected to criticism.

It is worthmentioning here that, in the light of the observations made by **S.R. Misra** (1995), in his project entitled "Identification of constraints in the Development of Inland Fish Cultivation with Special

Reference to West Bengal", the suggestions to overcome such problems seem more realistic. According to Misra, when canal irrigation is the source of water, then water should be released to fishponds at a very nominal rate. Again, since water is sometimes exhausted due to irrigation purposes, measures may be directed towards rain-fed farming on watershed basis. The run-off water collected into farm ponds should be used for irrigation.

T.A. Mammen (1983) points out that the farmers are not always interested to learn about the technical know how; that is, what he suggests that farmers should accept technical knowledge whenever it is available and in innovating new techniques, they should be open minded.

S.P. Sinha and **M.N. Jha** (1968) in their paper "Development in Fisheries" raised some relevant socio-economic problems. Their investigation in the context of co-operatives reveals the following :

- The influential and rich persons in Darbhanga do not want fishermen co-operative societies to grow at all. They do not want that the society should earn a substantial profit and give their members a stable profit. They assumed it would be a challenge to their socio-economic status.
- As most of the farmers are landless labourers, no financial help or subsidies or help of any kind were available to them. Rather, a stringent measures for recovery of the loans, result in disincentiveness of functioning of co-operatives on one side and on other side, encourage private trades to operate freely in their study area.

For the development of co-operative system, their suggestions are :

- Fisheries operating in a scattered way should be brought within co-operative system. Private trading, contract and auction system should be abolished.
- Fisherman Co-operative Societies should be aided with substantial financial help. Loans and grants are to be supplied to the members of the society at the right time and right occasion.

- Modern training and equipments should be supplied to them at a reasonable rate.

Difficulties in Co-operative Societies have been exposed by **G.S. Galgotia** (1968). His findings in many ways are similar to those of Jha and Sinha. But in addition to socio-economic analysis, he takes over the problems of marketing and storage. His findings can be summarized in the following way:

- The societies do not have sufficient accommodation facilities in the market for storage and hence they have to sell fresh fish within three or four hours and the members are compelled by circumstances to sell their produce to other dealers at much lower rates than the prevailing market prices.
- The societies have to face cut-throat competition from the private traders. At the time of heavy supply, the traders' joint efforts artificially push down the prices. Hence it becomes difficult for the members of the societies to meet up their cost and earn some minimum profits.

Galgotia has the same view like the previous investigators that government's discriminatory and unhelpful attitude are the main problems faced by the co-operatives. His suggestions were not much different than those of others. He alleged against government insincere attitude in supplying finance and credit. Similarly, the institutional credit for the processing units and transportation are not available up to the desired extent.

S. Ayyappan and **M. Krishna** (2004) in their article entitled "Fisheries Sector in India: Dimension of Development", published in *Indian Journal of Agricultural Economics* (IJAE) made an attempt to review the production potentialities and export possibilities of Indian fish production. They suggested different measures to meet the domestic requirements with an emphasis on earning reasonable amount of foreign exchange by way of extensive export measures.

In an article entitled "Institutional Structure for the Governance for Freshwater Aquaculture" in *Indian Journal of Agricultural Economics* (IJAE), **Marothia** (1984) examines the factors that affect the perfect utilization of multi-purpose common water bodies. He employed the contingent valuation method. On the basis of the data collected from common pool waterbody in Chattisgarh, he arrived at the conclusion that success in management of common pool water-bodies for fish culture does not depend exclusively on the technical-physical nature of the resource or the decision making arrangements or the behaviour of the members, but on the interrelationship of all the key parameters.

Ratna Reddy *et al.* (2004) in their article "Ecological and Economic Aspects of Shrimp Farming in Andhra Pradesh ", published in *Indian Journal of Agricultural Economics,* made an attempt to understand the economic incentives and ecological compulsions associated with shrimp farming. It was brought out that while the fish cultivation like shrimp farming has attracted a huge number of farmers in many states like Andhra Pradesh, its environmental consequences seem to be catching up fast. There is a need for judicious and sustainable management practices. It so happens in some states that the profits are vanishing as fast as they occur. In the process, the poor farmers are getting affected severely. The sector should be made responsible for the livelihood of the poor small and marginal farmers.

S.S. Shyam (2004) in their article "Export Performance of Indian Fisheries in the Context of Globalization", in *Indian Journal of Agricultural Economics* tried to articulate the export performance and potentials of inland fish products under the liberalized trade regime and explore the possible impacts of WTO agreements on the Indian Fisheries sector. He also identified the different factors like the commodity concentration, geographical concentration and the share of world import as the causal factors for the export instability.

Kazi M.B. Rahim and **Katar Singh** (1992) extensively studied the functioning of Central Fishermen's Co-operative Societies (CFCS) in West Bengal. Their findings highlight some mixed results about functioning of Co-operative Societies in West Bengal which may be quoted as follows :

> "The CFCS do not run profitably every year. It has to face negative net return from its business activities. The Primary Fishermen's Co-operative Societies (PFCS) do not avail any extra benefit other than the right of catching fish in the brackish water. The non-members too are not debarred from catching

the fish in the same water resources. Lack of administrative help is the main cause for these unlawful activities. The poor image about co-operatives in the mind of the members due to their experience is distinctly conveyed."

In an article entitled "A Study of Modelling of Household Demand for Fisheries in India", in *Indian Journal of Agricultural Economics,* **P. Kumar** and **M.M. Dey** (2004) tried to analyze the long-term changes that took place in fish consumption in India. An attempt has also been made to study the demand modelling on multistage budgeting framework and examined the demand elasticities (price and income) for fish. The responsiveness of demand for fish to changes in price and income (expenditure) varies for fish across income classes, regions and urbanization. Own price-elasticities of 'demand for fish is high' and has not shown any sign of decline during the last fifteen years.

Pradip K. Katiha *et al.* (2004) in their article entitled "Governance, Institutions and Policies for Fisheries of Floodplain Wetlands" in *Indian Journal of Agricultural Economics* presented the set of challenges for governance, institutions and policies towards development of fisheries. The study was based upon personal surveys and discussions with different stakeholders like fishers, Government agencies lessees, fishery co-operatives etc.

S.K. Chandan (2004) also in his article entitled "Institutional Management in Reservoir Fisheries", in *Indian Journal of Agricultural Economics* discussed, with the help of data collected from one hundred fishermen randomly selected from a reservoir fisheries in the Western Himalaya the socio-economic conditions of the fishermen, cost of fish capturing and marketing of fish through various channels in addition to the production and marketing management aspects.

In an article entitled "Enhancement and Culture-based Fisheries", **V.V. Sugunan** (2001) mentioned that considering the possible environmental impact and the socio-economic scenario of the country, it is not feasible to develop intensive aquaculture in the inland sub-sector, beyond certain limit. He also justified his analysis against the background of sustainable development of inland fishery resources. In another article entitled "Present Status of Inland Fishery

Resources in India and the Need for Their Enhancement" by **A.K. Ghosh** (CIFRI, Barrackpore, *Bulletin,* 2001) it is stated that enhancement of fish production from open waters has become a difficult proposition unless pursued on scientific lines. There is a need for holistic management policy, precisely for those inland waters which fall under common property resources. **P.K. Katiha** (2001) in one of his articles "Socio-economic Aspects of Culture-based Fisheries" (CIFRI, Barrackpore, *Bulletin,* 2001) observed that the interventions in the forms of the management and institutionalization of the waterbodies might be constrained by the wide variety of social considerations. Since the fishery development and management involve and affect primarily the fishers, it is necessary to consider the social aspects involving employment mobility, subsistence orientation of production and income distribution with regard to culture-based fisheries.

M. Sinha and **P.K. Katiha** (2002) in their article published in "Institutionalizing Common Property Resources", edited by **D.K. Marothia,** Concept Publishing Company, New Delhi, mentioned that, in addition to the activity of FFDA, aquaculture should be promoted with the necessary credit and technical support to fish farming agencies. Necessary changes in the lending norms of banks are a must for an easy flow of credit to the aquaculturists. They also pointed out that the preferential allotment of this water to production-oriented co-operatives with sufficient representation of fisheries would be an effective strategy for the development of the enterprises.

In an article entitled "The Bergram Majhipara Common Pool Fish Ponds" by **K. Singh** and **S. Bhattacharya** (1994) in "Managing Common Pool Resources: principles and Case Studies" by **K. Singh,** they observed in their study in case of West Bengal that ponds and tanks under common property regime had a low fish productivity. They clearly suggest that the government should promote this kind of water units on a co-operative basis and provide funds for the restoration of degraded common village properties, in addition to provision of technical guidance to fish farmers. In this context, **Marothia,** in his article "Village Ponds and Aquaculture Development: Is Conflict Inevitable?" stated that the multiple use of Common Village Ponds (CVPs) created conflicts and disorders among the fishers and village community. The study also revealed dissatisfaction over the caste-based preferential allotment of CVPs for aquaculture.

Jha (1997) in an article "Salient Ecological Features of the Chaurs and Mauns in Bihar and their Fisheries" in "Fisheries Enhancement of Small Reservoir and Floodplain Lakes in India" edited by **Sugunan** and **Sinha,** argued that the short-term leasing policy of the waterbodies by the fisheries department might affect productivity. He observed that in case of short duration of lease, lessees are not interested in investment in fisheries development. The fishers tend to catch out the entire stock of fish, resulting in further depletion of the fish stock. This amounts to convert these most fragile ecosystems into fishery deserts.

A review of the existing literature on this issue reveals that no systematic and comprehensive study has been made with regard to the problem that the technology developed which seems technically attractive has failed to produce substantial results in real world.

3

Methodology

"We must learn truths from facts."

—Mao De-Dong

The State of West Bengal is divided into the following six agro-climatic zones depending on soil characteristics and climatic conditions (Fig 3.1).

Coastal and Saline Region

Comprising Southern part of North and South 24-Parganas, Howrah and Midnapore districts. The soils are of tidal origin.

Old Alluvial Region

Comprising Mursidabad and Hooghly districts, part of Birbhum and Bankura districts, central part of the Burdwan and Midnapore. The soil in this region ranges from light to heavy texture according to the land situation. In general, soils are moderately well drained, low in organic matter and phosphate but high in potassium. Soil reaction ranges from acidic to neutral.

New Alluvial Region

Comprising Nadia district and parts of Malda, Mursidabad, West Dinajpur, Burdwan and Hooghly. In this region, soils are deep, moderately well drained and light texture, usually neutral to alkaline in reaction (pH level of 7.0 to 8.0). They are acidic to neutral (pH 6.0 to 7.0) in uplands. The soils have moderately high base saturation and medium to medium-low NPK status. Their calcium content is moderate to high.

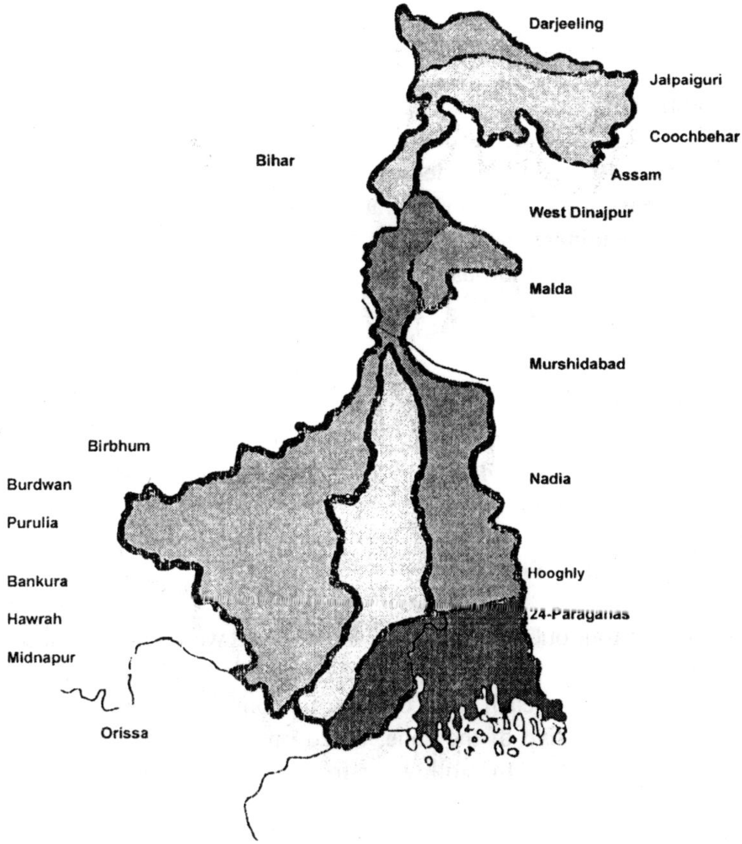

Fig 3.1: Agro Climatic Zones of West Bengal

Red and Laterite Soil Zone

Comprising Purulia district, parts of Birbhum, Burdwan, Bankura and part of Midnapur district with small pockets of Malda and West Dinajpur districts. The land is undulating with tiny rivulets and soils are highly eroded. They are mildly acidic (pH level of 6.2 to 6.8), poor in organic matter content, low in base status and have low nitrogen, calcium and available phosphate contents.

Terai Region

Comprising the plains of Jalpaiguri and Cooch Behar districts, Siliguri sub-division of Darjeeling district and Islampore sub-division of West Dinajpur district. Soils are deep and light textured, highly permeable and porous. Soils contain moderate level of organic matter without appreciable mineralization, highly acidic (pH level of 4.7 to 5.8) and low in bases, phosphorous, potash and micronutrients.

Hill Region

Comprising Darjeeling district excluding Siliguri sub-division and northern fringe of Jalpaiguri district. Soils in this region are generally shallow, coarse textured, high in organic matter content, low in bases and phosphate and acidic in nature.

Identification of Fish Farmers

Altogether four out of six zones in West Bengal were selected for the study. Two zones, viz. Hill region and Terai region had been excluded purposively. From each zone, one district considered comparatively more representative of the zone, based on agro-climatic situation, was selected. Thus the following four districts represented their respective zones :

Bankura (Old Alluvial region)

Birbhum (Red and Lateritic region)

Malda (New Alluvial region)

South 24-Parganas (Coastal and saline region)

A total of three blocks had been selected with equal probability without replacement from each district. Again a total of six villages had been selected from these three blocks according to equal probability without replacement. Village-wise list of pond-owners had been prepared from the mouza. The list of fish farmers with the number and area of ponds operated by them were prepared for all the villages selected. Then fish farmers were arranged in descending order on the basis of their water area (the aggregate of the water areas of different ponds held by an individual fish farmer). Then a sample of fifteen fish farmers in each identified village (or cluster of villages) were selected randomly, with the help of the random sample. Thus a sample of 90 fish farmers had been selected from each sample district. Finally, **a total of 360 (three hundred sixty) fish farmers were selected from four sample districts for the study**. **The unit of observation will be the fisherman's household**. Sampling design is shown in the following Fig. 3.2—

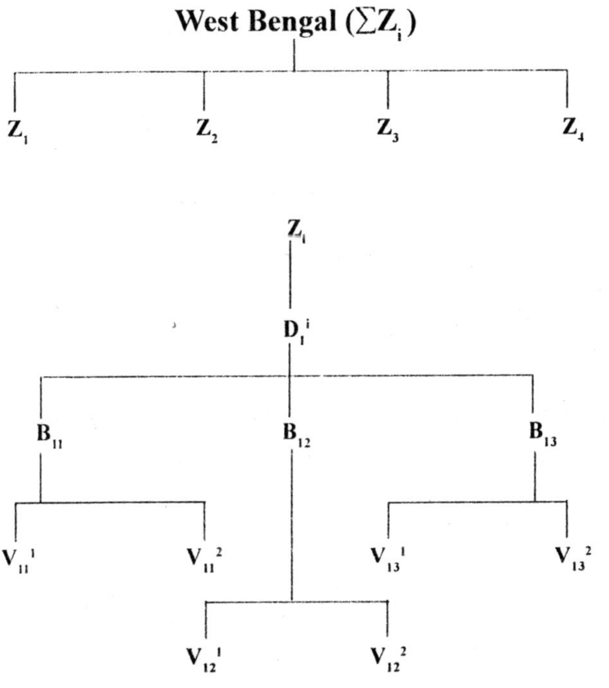

Fig. 3.2: Sampling Design

Where, $\mathbf{Z_i}$: ith zone
$\mathbf{D_1^i}$: district in representative zone
Bij : ith Block in jth district, i = 1, 2, 3; j = 1, 2, 3, 4
Vn : nth village in ith Block of jth district, n = 1, 2, 3, 4, 5.
Fû : ûth fish farmer in respective village, û = 1, 2,..........15.

Thus,

No. of fish farmers in a district = 6 × 15 = 90
Total sample size = 90 × 4 = 360

In this connection, it should be mentioned that in case of non-availability of 15 fish farmers in an identified nucleus village, two to four adjacent villages were identified to make a cluster of villages to have sufficient number of fish farmers for drawing its sample respondents. Names of blockwise identified villages (nucleus and associate villages) along with number of fish farmers, fish ponds and water area are presented in Exhibit 3.3.

Selected fish farmers were categorized into three groups on the basis of the size of ponds they held; that is,

(i) *Small pond owners:* less than 0.39 acre;
(ii) *Medium pond owners:* 0.4 acre to 4.99 acre; and
(iii) *Big pond owners:* 5.00 acre and above.

This exercise was carried for all the selected villages. Pond size groupwise number of sample fish farmers, number of ponds, water area and average size of water area are presented in the following Exhibit 3.1.

Exhibit 3.1: Size-wise Sample Fish Farmers along with Their Ponds, Water Area and Average Size

Category of Ponds	*No of Sample Farmers*	*No of Sample Ponds*	*Water Area (acre)*	*Average Size of water Area*
Small	261 (72.5)	298 (66.51)	83.52 (26.64)	0.32
Medium	88 (24.5)	127 (28.34)	158.4 (50.53)	1.8
Large	11 (3.0)	23 (5.15)	71.5 (22.81)	6.5
Total	360 (100.0)	448 (100.0)	313.42 (100.0)	0.87

Note: Figures in the parentheses indicate percentage to the respective totals.

The names of the selected blocks in each identified district are also presented in the following Exhibit :

Exhibit 3.2: Names of the Selected Blocks

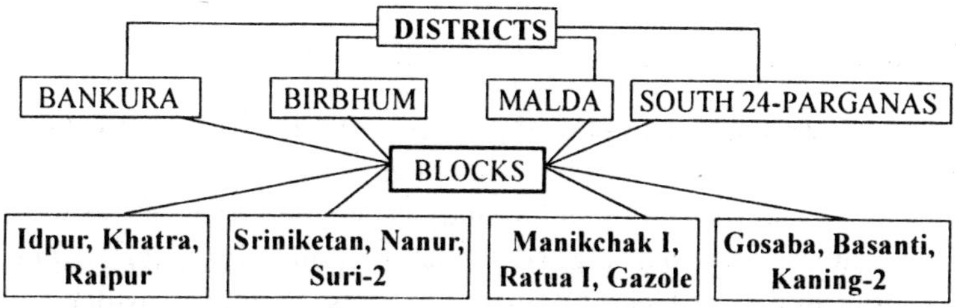

Selection of Nucleus Villages

The names of the villages of identified blocks along with their number of fishponds and water area were obtained from the respective Block Development Offices. Those villages are listed in the following Exhibit 3.3:

Exhibit 3.3: Blockwise Sample Villages Along with Number of Fish Farmers, Ponds and Water Area

Sl. No.	*Block*	*Sample Villages*	*No of Fish Farmers*	*No of Ponds*	*Water Area*
1.	Idpur	Lotihir, Basudevpur	30	38	26.6
2.	Khatra	Nakaijhuri, Ontha	30	36	24.8
3.	Raipur	Samsundarpur, Kalasole	30	37	25.7
4.	Sriniketan	Mahidapur, Ruppur	30	32	28.1
5.	Nanur	Lavpur, Langalhata	30	34	23.7
6.	Suri-2	Purandarpur, Aushgna	30	36	22.8
7.	Manikchak I	Kahala, Surjapur	30	37	28.4
8.	Ratua I	Kahala, Surjapur, Debupur	30	38	25.1
9.	Gazole	Kudbari, Birpara	30	39	24.3
10.	Gosaba	Vijaygarh, Rangaberia	30	40	32.9
11.	Basanti	Varatpur, Narayanpur	30	39	25.5
12.	Kaning-2	Rangaberia, Betberia	30	42	25.4

The data pertaining to the yield during the period, type and extent of water resources, physical characteristics, ownerships, adoption behaviour, costs incurred, marketing and other micro-details were culled by administering pre-designed structural questionnaires to all the fish farmers selected. Preliminary pilot survey was conducted in the villages in Sriniketan Block in Birbhum district to obtain first hand information about the problems[1]. Visits were also made to different District Fisheries Offices (Meen Bhavan) in order to elicit technological and institutional problems of inland fish cultivation of the respective districts through interaction with the officers.

The unique feature of the study is that the survey was confined to the areas of the four selected districts. In point of fact, it is pertinent to note that in the choice of the districts, certain considerations weighed most. In the first place, the impounded water units, mainly ponds and tanks, constitute a significant proportion of the total inland water resources; that is, heavy concentration of tanks and ponds in these districts. Secondly, in spite of the broad dissimilarities of soils, climate and other agro-climatic features among the selected districts, there were certain similarities in respect of different infrastructural facilities like availability of credit facilities, marketing etc. Finally, in all these districts, the Fish Farmers' Development Agency (FFDA) programme had been in operation.

Method of Enquiry

In this study, we have had a relatively open-ended and exploratory research design, which allows us to look at everything that appears to be relevant.[2] But as with all similar studies, the study suffers from certain general limitations. Firstly, the duration of the reference period is short. So, what one gets is a cross-sectional picture about an issue at a point of time. Secondly, the lack of specialized enumerations, the unwillingness or inability on the part of the fish farmers to provide detailed information, the widely scattered location of the production units and areas are some of the major problems. Incidentally, the maintenance of accounts by the fish farmers was not in evidence in the study area.[3] Respondents were interviewed by the project fellow. However, the random checking was also done by the Principal Investigator.

Period of Study

All the primary data obtained from fish farmers referred to the period 2003-2004.

Analytical Framework

Tabular analysis was done to arrive at relevant conclusions. In order to study the association between output and various inputs used, the following model of Cobb-Douglas production function was used.

$$Y = Ax_1^{b_1} x_2^{b_2} x_3^{b_3} x_4^{b_4} x_5^{b_5} e^u$$

Where,

Y	:	Value of gross fish production
A	:	Intercept (constant)
x_1	:	Fertilizer and manure (Rs./kg)
x_2	:	Mahua oil Cake (Rs./kg)
x_3	:	Fish feed (Rs./kg)
x_4	:	Fingerling (Rs./kg)
x_5	:	Labour (Rs./kg)
e^u	:	Error term
b_i	:	Regression coefficients of respective variables

To examine the productivity of different inputs used in fish production, marginal value productivities of inputs were estimated by the following formula:

$$MVP_{x_i} = \frac{b_i Y}{X_i}$$

Y = Geometric mean of the gross fish production;

X_i = ith input;
x_i = geometric mean of the ith input; and
b_i = Elasticity of production of the ith input

Tests of association like chi-square test, correlation etc. were used. Bar diagram, pie charts and other descriptive statistics were also used.

NOTES

1. After making a detailed pilot survey, some modifications in the questionnaire were made.
2. Often, the investigator had to pay several visits to an individual to elicit detailed information from its selected respondents on various items of interest through probe questions.
3. For example, it is true that estimates of catch with a reasonable degree of precision can be obtained at a very high cost. For inland waters, the areas are generally scattered and unorganized with irregular catch operations.

4

Structural and Operational Features of Sample Tanks and Ponds

> If India's agrarian history demonstrates anything, it is that doing and saying nothing is preferable to taking small steps timidly and slowly.
>
> —*Daniel Thorner*

In this chapter, attempt has been made to examine some of the important aspects of structural and operational features of the ponds and tanks like size of the water units, types of possession, size productivity relationship etc.

Ponds and tanks vary in size and, therefore, a study of the size distribution of the sample water units would be relevant for economic analysis. The ponds and tanks are grouped into three different size classes. Exhibit 4.1 gives the distribution of the number of water units and the area (percentage-wise) for these three size classes (as defined earlier in chapter three).

It is evident from Exhibit 4.1 that average size of the water units (under survey) is approximately 0.87 acres. It is also observed that more than 88.5 per cent of the water units with more than 40 per cent of the total water area are of size less than the average size. Water units of size 5.00 acres and above constitute only 3 per cent of the total water units but dominate a sizeable proportion (22.8 per cent) of total area. And water units of size between 0.40 to 4.99 acres constitute 50.6 per cent of the water area with an average size of 1.8 acres.[1] In Fig. 4.1, the Lorenz curve clearly depicts the heavy concentration of water units into small size.

Exhibit 4.1: Size Class-wise Distribution of the Number and the Area of Currently Operated Water Units in 2004

Size class of Water Units (acres)	*No of Ponds*	*Percentage of the number of Units*	*Percentage of the Area of Water Units*	*Area per Unit (acres)*
Up to 0.39 (Small)	298	66.52	15.4	0.32
0.40–4.99 (Medium)	127	28.34	56.9	1.80
5.00 and above (Large)	23	5.14	27.7	6.50
Total	448	100.00	100.0	0.87

Source: Field Survey, 2004.

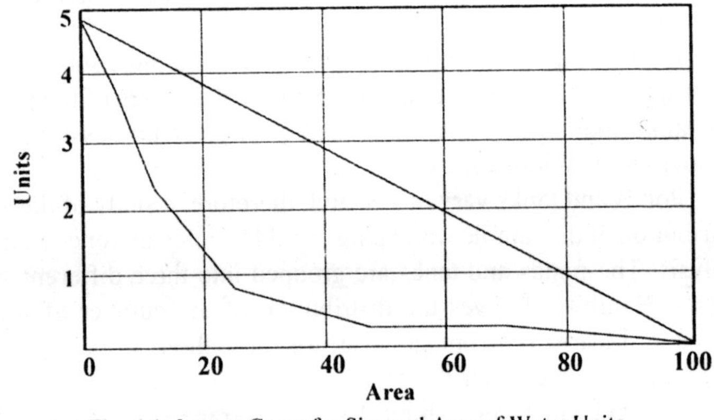

Fig. 4.1: Lorenz Curve for Size and Area of Water Units

Size group-wise ownership and extent of proprietorship of ponds and tanks were estimated and shown in Exhibits 4.2 and 4.3. It was found that in the study area, the water units (ponds and tanks) were either owned or leased in through different sources. The water units were operated either as solo proprietorship or by the multiple of operators as partners. From the following Exhibit 4.2 it is evident that 294 (65.62 per cent) of the total sample water units were cultivated by the owner-operators; whereas 154 units (34.38 per cent) of the total sample units were leased in. A significant proportion of the owned and cultivated by the owners were found in the cases of small and medium size of water units, i.e. 70.13 per cent and 61.42 per cent of the total sample of the respective size-groups. Only a small fraction of the large-sized water units are operated by the owners, i.e. 30.44 per cent; whereas, 16 units or 69.59 per cent of the total units under large size group were found leased out.

During the course of investigation, it was observed that the preponderance of the owner-operators and relatively less importance attached to leasing out system might be accounted for by two reasons. One, with the advent of new technology of composite intensive fish culture, self-cultivation is being considered by the owners a paying proposition; and two, because of a series of tenancy legislations in recent years giving more rights to the tenants, the owners have become more apprehensive of losing the occupancy rights in their leased out water units.

Exhibit 4.2: Size Group-wise Distribution of Ownership of Water Units

Category of Ponds	*No of Ponds*	*Ownership Status*	
		Owned	*Leased*
Small	298 (100.00)	209 (70.13)	89 (29.87)
Medium	127 (100.00)	78 (61.42)	49 (38.49)
Large	23 (100.00)	7 (30.44)	16 (69.56)
Overall	448 (100.00)	294 (65.68)	154 (34.38)

Note: Percentages in parentheses indicate the respective percentages.
Source: Field Survey, 2004.

Exhibit 4.3: Size Group-wise Distribution of Proprietorship of Water Units

Category of Ponds	*No of Ponds*	*Extent of Proprietorship*		
		Sole	*Only 2 Partners*	*More than Two Partners*
Small	298 (100.00)	93 (31.21)	178 (59.73)	27 (9.06)
Medium	127 (100.00)	32 (25.19)	65 (51.18)	30 (23.63)
Large	23 (100.00)	6 (26.09)	10 (43.48)	7 (30.43)
Overall	448 (100.00)	131 (29.24)	253 (56.47)	64 (14.29)

Source: Field Survey, 2004.

It can be mentioned that water units in the study area were found to be usually operated by different agencies like individuals, either singly or jointly, co-operatives, government or local bodies. Joint private operations and solo private operations were found to be the major agencies. From Exhibit 4.3, it is evident that quite a significant proportion of the water bodies were found to be operated by more than two partners, particularly in the cases of large and medium sized water units (23.63 per cent and 30.43 per cent). Comparatively, a large number of water units (29.24 per cent) were operated by solo pond owners. It is important to note that quite a good proportion of water units are cultivated by a large number of operators. Unlike other States in India, it had been found in our study area, in West Bengal, the preponderance of multiple ownerships of water units, irrespective of size-classes.

As we have already mentioned in Exhibit 4.2 that in our study area, 34.38 per cent of all the ponds surveyed were leased in. Three distinct sources were mainly identified with which lease agreements were made – they were private owners, public bodies and others like Panchayat, Zilla Parishad, trust or institutions like schools etc. The duration of lease was of various time periods ranging from one year to ten years.

It is evident from the Exhibit 4.4 that 118 out of 154 water units leased in (i.e. 76.62 per cent) were found to be obtained from private sources; whereas, 22 out of 154 (i.e. 14.29 per cent) of the total water units were found to be leased in from government or public bodies. Interestingly enough, in West Bengal, most of the water units of all size-categories were leased in from the private sources. 76.40, 79.59 and 68.75 per cent of the water units belonging to small, medium and large size-groups respectively were leased in from the private sources.

Exhibit 4.4: Size Group-wise Leased in Ponds along with their Leasing Sources

Category of Ponds	*Lease in Ponds*	*Leasing Sources*		
		Private	*Public*	*Others*
Small	89 (100.00)	68 (76.40)	12 (13.48)	9 (10.11)
Medium	49 (100.00)	39 (79.59)	6 (12.25)	4 (8.16)
Large	16 (100.00)	11 (68.75)	4 (25.00)	1 (6.25)
Overall	154 (100.00)	118 (76.62)	22 (14.29)	14 (9.09)

Source: Field Survey, 2004.

It may further be observed from Exhibit 4.5 that majority of the water units (i.e. 115 out of 154 ponds or 76.68 per cent) were leased in for a short duration, i.e. for one year; whereas 32 and 7 water units out of 154 water units surveyed, i.e. 20.78 per cent and 4.54 per cent respectively were found leased in for a period of two years or more than two years respectively. Pond size-wise analysis of the duration of the lease period reveals that majority of the water units of small and medium size-groups (79.78 per cent and 77.55 per cent of the total water units of the corresponding size-groups respectively) were leased in for a short period of one year only. However, the duration of lease period for more than four years was found conspicuous by its absence in the case of medium sized water units. Only one-fourth of the large ponds (25 per cent) were leased in for more than four years. 37.50 per cent of the large water units were leased in for only one year and also for a period of two to four years.

Exhibit 4.5: Size Group-wise Leased in Ponds along with their Lease Duration

Category of Ponds	*Lease in Ponds*	*Lease Duration*		
		One Year	*2–4 Years*	*More than 4 Years*
Small	89 (100.00)	71 (79.78)	15 (16.85)	3 (3.37)
Medium	49 (100.00)	38 (77.55)	11 (22.45)	— (0.00)
Large	16 (100.00)	6 (37.50)	6 (37.50)	4 (25.00)
Overall	154 (100.00)	115 (74.68)	32 (20.78)	7 (4.54)

Source: Field Survey, 2004.

The figures in Exhibit 4.6 bring out the fact that most of the water units (52 per cent) were acquired through inheritance and only 10.5 per cent of total water units were purchased. The share of the construction of new water units in the total number of water units was 3.12, whereas 34.38 per cent of the total units were acquired through leasing arrangements. It should be stated that it is the inheritance factor which is mainly responsible for fragmentation of ownership (that is, multiple ownership).[2]

Exhibit 4.6: Percentage Distribution of Water Units by Means of Acquisition

Means of Acquisition	*Percentage of the Number of Water Units (of 448 water units)*
Inheritance	52.0
Purchase	10.5
Construction of new water units	3.12
Lease	34.38
Overall	100.0

Source: Field Survey, 2004.

The range of uses of water units is very large and includes uses like irrigation, duck rearing, cattle watering, pisciculture field drainage and also for crop production for crop for both cattle and human being. The following Exhibit 4.7 demonstrated that almost all water units are used for multipurpose with a wide variety of combination of uses. 'Household uses' in our case include bathing, dishwashing, cloth-washing, supply of water for cooking etc. 58 per cent of the water units are used for household purpose, 61 per cent of the ponds and tanks are also used for irrigation purposes.[3] There are inlet and outlet channels indicating that the units may be used for drainage and irrigation also.

It can also be pointed out that the water units sometimes provide both water and space for growing vegetables. Dried out parts of the ponds are sometimes used for winter (Boro) paddy cultivation—either for nursery seedbeds or for transplanting the paddy itself.

Exhibit 4.7: Use of Water Units

Household Purposes	*Fish Production*	*Field Irrigation*	*Duck rearing, cattle watering etc.*	*Percentage Reporting Combination*
1	*2*	*3*	*4*	*5*
	*			25
	*	*		19
*			*	10
*	*			20
*		*	*	12
	*	*	*	8
	*		*	6
58	68	61	45	100

Note: Percentages are reporting this use.
Source: Field Survey, 2004.

In point of facts, it may be stated that there is no systematic relationship between size and yield of ponds. The correlation coefficient is found statistically insignificant. During the course of interview, it was observed that some ponds owned and operated by an enterprising fish farmers located in an ideal situation may yield higher fish production, irrespective of the size of water units. Water units of small size enjoy sometime the benefits of closes supervision, intimate monitoring and sincere regulation by the family members themselves. On the other hand, large sized ponds may also have the privilege of initiative and care likely to be taken up by enterprising commercially motivated market oriented fish farmers. Some amounts of business elements sometimes play some role in its operation.

Another interesting feature which might have wider policy implications is that the larger the number of shareholders in a water unit, the lower the level of yield. In Exhibit 4.8 it was shown that there has been a trend towards decline in yields in water units as the number of shareholders increased. Water units cultivated by the solo operator has the yield of 608 kg per acre; whereas, the units owned by, for example, 5 or more shareholders, had the yield of 427 kg per acre. A large number of water units (ponds and tanks) under multiple ownership are found to be weed-infected and semi-derelict. Because of the recurring deposition of the silt, the bottoms of these water units often dry up during the summer and the water units overspill during the monsoon. Immediate desilting of the tanks, eradication of the excess vegetation and water hyacinths are required to be made in order to provide appropriate conditions for fish cultivations.

Exhibit 4.8: Distribution of Yield by Number of Shareholders

Number of shareholders		*Yield per Acre (kg)*
Single	1	608
Joint:	2	510
	3	540
	4	408
	5 and above	427

Source: Field Survey, 2004.

In the case of jointly owned water units, it was observed that the shareholders of unequal economic conditions and sometimes absentee shareholders were, more often than not, found reluctant to take initiatives for renovating water bodies, as their shares of benefits are small; on the contrary, it was observed that in all these cases, there were instances of creating problems by shareholders. Sometimes familial conflicts among the shareholders have hindered the adoption of modern fish culture practices and improvement of ponds and tanks. Different shades of opinion arose among them. All in all, sometime, age-old water units are eventually remaining in precarious conditions without any efforts for proper maintenance.[4]

Exhibit 4.9: Yield per Acre by Duration of Lease Period

Duration of lease period	*Yield per acre (kg)*
On annual basis	570
2 years	590
3 years	575
4 years	640
10 years	585

Source: Field Survey, 2004.

From the Exhibit 4.9, it can be safely asserted that longer the duration of lease period, the higher the yield. The correlation coefficient between yield and duration of lease period is found to be 0.47. This is statistically significant at 10 per cent level of significance. The long duration of lease period provides the fish cultivators incentives and generates a kind of mind-set to undertake different kinds of investment, short run and long run, through adoption of different culture practices. All this might play a positive role in augmenting fish production in a water unit.

NOTES

1. A study carried out by Central Inland Fisheries Research Institute (CIFRI, Barrackpore) during 1978-81 in 24-Parganas, revealed that ponds of size less than 0.4 hectare are predominant with an average size of 0.1 hectare, quoted in *Scope for Scientific and Commercial Fish Farming in West Bengal* by Sri A.V. Natarajan.
2. Hindu laws of inheritance, "Mitaksara" and "Dayabhaga", the dominance of which continued uninterrupted in the countryside played a very

important role in subdivision and fragmentation of lands and ponds in India. This is eloquently explained in Pronoy Roy's "Subdivision of Land Soil Quality Biases: A Study of Property Rights in Indian Agriculture" (*Indian Journal of Agricultural Economics*, June 1983, p. 193). It is sometimes stated that zamindari abolition in some States has resulted in multiple ownership.

3. Some ponds and tanks are seasonal irrigational tanks only. The demand for water from these ponds for irrigation purpose is heavy, resulting, in many cases, in complete draw-down. In these water units, it is only a put and take fishery because of limited period of water storage.
   A study carried out by CIFRI during 1978-81 in 24-Parganas revealed that only 33 per cent of the water area covered by ponds is used exclusively for fish cultivation, while 64 per cent of ponds (by area) are used for different purposes including fish cultivation, while three per cent are remaining fallow.
4. It is important to note that the problem of fish culture does not stem from multiple ownerships as such, or even from unequal ownership shares; but from multiple ownership by persons 'forced together by inheritance' or some other reasons.

5

Tenancy Relations and Resource Use Efficiency in Pisciculture

The fish farmers either owned or took on lease the waterbodies (tanks and ponds). In the whole of India, 31 per cent of the ponds were owned and 69 per cent were on lease (IIM, 1990). Of course, the proportion varied across the states. It has been observed that in states like Rajasthan, Karnataka, Himachal Pradesh, there did not exist any ownership ponds; whereas in states like West Bengal, Assam, Orissa we witnessed the predominance of ownership ponds. Most of the fishponds were owned by government agencies like reservoirs, irrigation revenue, government agencies, fisheries department or village Panchayats. These water bodies were common property resources.

In West Bengal, unlike other states in India, majority of the tanks and ponds are owned by private parties. A significant proportion of these waterbodies are found to be leased out to different beneficiaries. The beneficiaries who took the ponds on lease were generally of three types: (a) single individual, (b) two or more individuals and (c) members of co-operatives. The criteria for selecting the leases vary. Some are leased out to working fishermen or persons from fishermen community or unemployed or economically backward rural youth interested in fish farming; some are also sometimes leased out to co-operatives organized by the enterprising individuals with knowledge in pisciculture. On the strength of field data, we have also reason to believe that there are also some cases where the pond owners are interested in leasing out to those lessees who have already proved trustworthy. The number of pond owners covered in our study is 120 in the district of Murshidabad. The total number of ponds studied is 188. Of them, 112 (59.57 per cent) were leased out to individuals.

It was also observed that different kinds of ownership of water units called for different kinds of procedures for leasing.[1] Different owners followed different leasing periods and different criteria for selecting tenants. Absence of uniform leasing period is an important aspect. The uncertainty attached to short duration lease generally dampens the motivations of the lessees. There is no guarantee that the existing lessee would get the same ponds and under the same conditions in the next year also. Automatically, he is wary of taking up productive investment. The reasons for leasing out ponds may be :

- Some have small fragments of ponds which are found convenient to lease out.
- A few pond owners do not have adequate manpower or skill or capital and the ability to undertake fish cultivation; and hence are unable to pursue fish cultivation by themselves.
- A few others have other occupations which could be profitably pursued otherwise.
- All these apart, pond owners belonging to the upper caste have the tendency to shun the harsh life of fish cultivation and hence, the tendency to lease out.
- Some pond owners who are sometimes in dire need of cash tend to lease out their ponds to some enterprising relatively affluent lessors.

No other State in India have such a bewildering variety of ponds leasing system as in West Bengal.[2] The modern pisciculture has become gradually more capital intensive in nature and profitable. This has made taking the ponds on lease more attractive. Different types of tenancy contracts co-exist in our study area, as in most other parts of West Bengal. At a general level, two forms of tenancy contracts prevail. They are (1) share contract and (2) fixed rent contract.[3]

Share Contract

This is when rent takes the form of share of the harvest.[4] The amount of rent is a matter of mutual agreement between the pond owners and the tenants. Pond owners and tenants choose among a limited number of standardized contracts. It largely depends upon the location, quality, size of ponds concerned etc.

In our study area, the prevalence of oral lease system mainly in privately owned ponds is observed. Crop sharing tenancy is more prevalent among the small tenants. This is because small tenants find it more difficult to pay the cash rent.

Under crop sharing tenancy system, it was the perception that the tenant would have to part with a part of the surplus, created by his own labour. Even when the hired labour is employed, generally the excess labour costs are borne by the tenants (even though there are instances where the costs for hired labour are equally shared by the tenants and the pond owners).

Fixed Rent (in Cash or in Kind)

Rent is paid in cash in most cases at the beginning of the season. Tenants are solely responsible for all the aspects of cultivation. Fixed rent in kind has the same features as the fixed rent in cash except that the rent mutually agreed is paid sometimes at the time of final harvest along with some occasional catch as and when both the tenants and the pond owners mutually desire.

The pond owners are sometimes found an important source of credit to the tenants. Cost sharing is a standard part of the tenancy arrangement. The tenants are found to provide traditional labour intensive techniques and inputs like farmyard manure, mahua oil cake, rice bran, family labour; whereas, modern capital intensive inputs like fingerlings, fertilizers etc. are either wholly or partly provided by the pond owners. The reasons for sharing capital-intensive inputs are quite straightforward on efficiency ground. There are efficiency gains from input sharing since it gives both pond owners and tenants an incentive to equal cost and marginal revenue. Cost sharing has emerged as an adaptation to the requirements of modern pisciculture.[5]

Incidence of production loan is also not insignificant. The pond owners give quite often credit to the tenants and participate in decision-making about the units of those inputs and in general, takes a lot of interest in productive investment on the leased out ponds. Of course, a few cases were observed where the terms and conditions of contracts in tenancy and credit transactions or labour are interrelated. The implications in the relevant factor markets might get re-enforced by such inter-linkages.[6] The pond-owners may rob his tenants of his freedom in decision-making.

In 20 per cent of the cases of the reporting tenants in the sample the pond owners gave advances to the tenants to meet his needs for the purchase of inputs like fingerlings etc. In the rest 80 per cent cases, advances were given without any interest.

The information received would have us believe that the most distinctive aspect in the lease market is the rising incidence of fixed rent tenancy and the entry of some middle-class enterprising individuals as lessors-in. This emergence and growth is a reflection of development of capitalist tenancy as distinguished from subsistence tenancy. This has, of course, reduced the growing tendency for reduction in the access of small/marginal fish farmers.

It is perceived that tenants under cash rents tenancy arrangements have the freedom to utilize the ponds with the objective of production. Cash rent tenancy (fixed) is more profitable; because the whole of economic surplus, after meeting cultivating expenses, is enjoyed by the cultivators. That is why, more resourceful tenants prefer to take ponds on cash rents. Of course, it has been found that sharecropping provides both conflicts and co-operations between pond owners and tenants. It also involved joint provision of many inputs. Exhibit 5.1 summarizes the terms and conditions of different tenancy contracts.

Exhibit 5.1: Types of Tenancy Contracts

Contract	*Proportion of leased in area under specific contracts (%)*	*Tenants share in output (%)*	*Tenants share in output (%)*			*Other features, if any*
			Labour	*Inputs like oil cake, rice bran etc.*	*Cash inputs like fish seed, fertilizers etc.*	
1	2	3	4	5	6	7
Share Contracts	40	50	100	100	100	Cost share shared at the time of occurrence
	15	33.3	100	100	100	
	10	25	100	00	00	
Fixed rent contracts						Rent is paid sometimes at the beginning of the season or sometimes at different points of time
Cash rent	20	100	100	100	100	
Kind rent	15	100	100	100	100	

Source: The Field Survey, 2004.

The Exhibit gives an impression that the highest concentration of ponds is governed by sharecropping. The urban households (or absentee landlords) and the non-labour institutions are generally found to prefer fixed rents. In about 40 per cent of the ponds, output is shared in different proportions between the pond owners and the tenants. Output is shared equally when all the input costs like labour, seed fertilizer etc. are entirely provided by the tenants. There are reductions in the tenant's share to 33.3 per cent or 25 per cent of the total produce when the variable input costs partly or wholly are borne by the pond owners. The pattern of association between crop sharing and cost indicates a tendency towards pond owners sharing costs only when they are receiving a larger share of the crop. There are also some cases where instead of one tenant, and one landowner, there are more than one tenants or owners for one pond owner or one tenants with identical obligations and entitlements. This system generates mainly because of the multiplicity of ownership. In most cases, however, each co-owner or co-tenant provides usually proportionate share of inputs and received proportionate output.

Resource Use Efficiency

There is no denying the fact that inland fisheries sector in India is being utilized to provide employment and to improve the socio-economic status of the inland fishing community. This community is largely of subsistence nature. It is again sometimes observed that the sub-optimal input use has led to a very high cost of fish production and resultant low returns.[7] No serious attempts have so far been made to identify the key input factors and their contribution to fish production. Hence, there is a need for developing an optimal resource management strategy and the appraisal for resource use efficiency.

Production of fish depends upon various factors. Through production function analysis, an attempt has been made here to set up a relationship between production of fish and different inputs like fish seeds, fertilizers, supplementary feed, manures, labour etc. Following is the general form of production function (Cobb-Douglas type of production function) which is employed to estimate production elasticities.

$$Y = a \Sigma b_i x_i u_i$$

$$\text{or, } Y = Ax_1^{b_1} x_2^{b_2} x_3^{b_3} x_4^{b_4} x_5^{b_5} e^u$$

Where,

Y : Value of gross fish production
A : Intercept (constant)
x_1 : Mahua oil cake, fertilizer, lime, manure etc. (Rs./kg)
x_2 : Fingerling (Rs./kg)
x_3 : Fish feed (Rs./kg)
x_4 : Labour (Rs./acre)
e^u : Error term
$b_1 - b_4$: Regression coefficients of respective variables.

The study is based upon primary data. Data pertaining to output, cost on labour, fish feed, fertilizer etc. were elicited from 90 fish farmers selected for the study from a district of West Bengal by administering the questionnaire designed for the purpose. The analysis is based on cross-sectional pond-wise per acre value of output (dependent variable, Y) and the values of all the four independent variables explained above. The marginal value product (MVP) of an input is given by

$$MVP_{X_i} = \frac{b_i Y}{X_i}$$

Y : Geometric mean of the gross fish production
X_i : i^{th} input
P_y : Price of the output (Rs./kg) and
b_i : Regression co-efficient of the ith input

The calculated F-value is 8.4 which is greater than the table value (F = 3.87) and significant at five per cent level of probability. It clearly suggests that the independent variables included in the study affect fish production significantly. The estimated value of R^2 is 0.68. This indicates that 68 per cent of the total variation of fish production (Y) is explained by four inputs which have been included in the analysis.

The sum of the production elasticities indicate the nature of the returns to scale. The sum of the elasticities of all the four variables under analysis is 2.09. This indicates that the increase in one per cent of the variable will result in 209 per cent increase in the fish production. Hence, in our case, increasing returns to scale seems to exist in the production of sample ponds. It is also evident from the Exhibit that individual production elasticities of manure, lime and fertilizers (X_1), fingerlings (X_2), feed (X_3) included in the study emerge not only positive but also significant at 5 per cent level of probability. These variables influence production significantly. It can also be stated that when other inputs remain constant, an increase in one per cent investment in a single input like fingerling (or fish feed) will bring about an increase in fish production to an extent by 80 per cent. The similar conclusion can be drawn with regard to other inputs.

In fact, the highest expected increase in fish production has been noticed due to increase in investments in fish seeds (fingerlings); because increase in investment on this input may help fish farmers to utilize better quality of fish seed which is the most critical input. The coefficient of labour is found to be 0.09 which is statistically insignificant. This implies that an increase in employment of labour without corresponding increase in complementary inputs like better quality of fish seed, or manure or fertilizer, may not necessarily generate any increase in fish production. This suggests that under the existing production conditions, there is a need for increase in investment on material inputs along with labour.

Again, it is also clear from Exhibit 5.2 that Marginal Value Productivities (MVPs) of all the inputs are found to be positive. In all the cases, the MVPs have values more than one. This explains that an investment of Rupee one on any input under analysis is likely to increase fish production by more than the value of Re. one. In other words, significantly, an investment of Re. one in fish seed is likely to generate an additional income to the extent of Rs. 5.2 through fish production. Of course, this largely depends also upon the price of fish. All these observations suggest that there exists ample scope for improvement of fish production through appropriate input allocation. It can be noted that every water units have a maximum capacity to support certain level of fish stock (carrying capacity), while stocking in a water unit, care should also be taken to see that proportionate stocking with other inputs are made for optimal utilization of all the inputs.

The estimated production function for the sample studies

provides approximate rather than very precise estimates of input-output relationship in the production of fish under study.

Exhibit 5.2: Production Coefficients, Standard Errors and MVPs of Various Inputs

Variables	*Production Elasticities*	*MVPs (Rs.)*
Manure, lime and fertilizers (X_1)	0.7* (0.07)	4.8
Fingerlings (X_2)	0.8* (0.19)	5.2
Feed (X_3)	0.5** (0.10)	3.1
Labour (X_4)	0.09*** (0.35)	2.1

Figures in the parentheses indicate standard error.
* significant at 5 per cent level of probability.
** significant at 10 per cent level of probability.
*** not significant.
$R^2 = 0.68$; F-value = 8.4 (significant at 5 per cent level of probability)

NOTES

1. Tenants continue to cultivate same ponds without any official record of his cultivating possession over the ponds (i.e. concealed tenancy). Oral lease is a word of mouth agreement between the lessor and the lessee. In order to evade tenancy legislations, the system of oral leasing is sometimes practised.
2. The term "reverse tenancy" refers to the situation where large pond owners are leasing in ponds from smaller ones. The phenomenon is not that significant in the case of fish cultivation. In fact, in agriculture this tendency prevails because the land owners are to make full use of some indivisible inputs like pump sets, threshers etc. and hence take recourse to reverse tenancy. But in pisciculture, there does not exist any input indivisible. Hence, the tendency is found to be low.
3. One may encounter some other forms of contract which do not fit in easily with either category.
4. It is sometimes argued that crop sharing leads to misallocation of resources. It is (not) possible to accept that crop sharing arrangement leads to more efficient allocation of resources. Risk is shared between the pond owners and the tenants. In pisciculture, which involves high degree of uncertainty, the incidence of sharecropping is prominent.
5. Like agriculture, pond owners and tenants cannot be described as sharply differentiated class. With conflicting interests, tenants and pond owners

are found often in the study area involved in a "co-operative conflict" with many "congruent interests" (like the quantity of fish seeds, fertilizer use etc.) and also many "conflicting interests" (like division of the harvest between the two parties) (See Sen, 1990).

6. There are very few cases where the terms and conditions in tenancy and credit transactions or labour are found to be inter-related. The implications are that the imperfections in different factor and commodity markets might get reinforced by such interlocking. The pond owner may have the opportunity of robbing the tenants of the freedom of the tenants in decision-making.
7. Values of the simple correlation coefficients between the independent variables are found to be smaller than the values of R^2. So, by this test, the estimated equation may be taken as not marred by the problem of multicollinearity. See Lawrence R. Klein, *An Introduction to Econometrics,* Princeton-Hall (Eastern Economy), New Delhi, 1965, p. 144.

6

Constraints for Adoption of Package of Practices

"Do not believe in traditions, merely because they have been handed down by many generations and in many places; do not believe in anything because it is spoken of by many; do not believe because the written statement of some old sage has produced.....After observation and analysis, when it agrees with reason and is conducive to the good and benefit of one and all, then accept and live up to it." —The Buddha in Kamasutra, Anguttara Nikaya

There is no doubt that the fish production has been a traditional enterprise in our study area. In the past it was practised thoroughly with traditional method. But during the last few years, precisely after launching of FFDA programmes, steps are used to be taken by the fish farmers to adopt modern culture practices. It is of interest now to examine the response of operators to the package of intensive culture practices over time. In this study, we define adopter-operators as those operators who adopt the practices like pre-stocking, stocking, input applications like lime, toxicants, manures, fertilizers, application of supplementary fish feeds, harvesting etc.

Exhibit 6.1 below provides an overview of the trend with regard to adoption of practices in the districts by size classes of operators, trained and untrained and receiving and not receiving assistance of any kind from different sources. The percentage of operators of the water units of different size-classes who received training and assistance adopting the package of practices are found to be 37.22 per cent and 38.89 per cent in 2001 and 2004 respectively, indicating thereby that the pace of adoption is not very satisfactory. As against this, only 17.5 per cent and 15.83 per cent of the operators (untrained and did not receive government assistance) have so far used package of practices in 2001 and 2004 respectively. Again, it is important to note that there has been a marginal decline in adoption behaviour of those farmers who are not trained but received assistance. In other words, the percentage of the fish farmers, adopting culture practices decreases from 3.89 per cent to 3.61 per cent from 2001 to 2004. However, no change was witnessed in the cases of those fish farmers who neither were trained nor received any kind of assistance.

Exhibit 6.1: Distribution of Different Types of Operators Adopting Modern Culture Practices by Size-classes in 2001 and 2004

Size class of Operators	*Distribution of sample farmers*							
	Category A		*Category B*		*Category C*		*Category D*	
	2004	*2001*	*2004*	*2001*	*2004*	*2001*	*2004*	*2001*
Small	94 (36.01)	90 (34.48)	40 (16.85)	47 (18.00)	9 (3.44)	8 (3.06)	117 (44.82)	116 (44.44)
Medium	39 (44.31)	38 (40.91)	15 (17.04)	14 (15.90)	4 (4.54)	6 (6.81)	30 (34.09)	30 (34.09)
Large	7 (63.63)	6 (54.54)	2 (18.18)	2 (18.18)	0 (0.00)	0 (0.00)	2 (18.18)	3 (27.27)
Over all	140 (38.89)	134 (37.22)	57 (15.83)	63 (17.50)	13 (3.61)	14 (3.89)	149 (41.38)	149 (41.38)

Figures in the parentheses indicate the respective percentages.

Note: Category A - Adopters who received government assistance and trained.
Category B - Adopters who are trained but not received financial assistance.
Category C - Adopters who are not trained but received the financial assistance.
Category D - Adopters who are neither trained nor received the financial assistance.

Source: Field Survey, 2004.

The Exhibit also makes an interesting reading as far as a comparison of the level of adoption by size-class is concerned. An upward movement of the adoption of modern culture practices was also registered by those fish farmers, belonging to small size-class, trained and received assistance from 34.48 per cent to 36.01 per cent to all the small fish farmers surveyed during the year 2001 and 2004. Interestingly enough, there has been also an upward trend in the adoption behaviour of those small fish farmers who received neither training nor assistance of any kind, i.e. an increase from 44.48 per cent to 44.42 per cent during the same period.

Here also, we employ test of association to examine whether training and Government loan-subsidy programme have got any association with modern culture practices. Also quite naturally the degree of association between the attributes – (i) training and adoption, (ii) financial assistance and adoption – are found to be + 0.28 and + 0.32 respectively.

Certain interesting observations are in order :

One, the package of practices appealed to those fish farmers who are trained and reasonably backed by financial assistance by way of loan/grant with or without any provision of subsidy. **Two**, training alone by itself may not necessarily stimulate all the fish farmers, particularly small and medium sized farmers to go in for modern culture practices. Other complementary factors like financial assistance, may act as a positive factor. **Three**, over a period of time, the fish farmers, particularly the medium sized farmers, tend to practise the modern culture practices through receiving the information/knowledge from the other farmers. Sometimes, in our study area, it was observed that even the small farmers are found to become enthusiastic to internalize the externalities of efficient production system taking place elsewhere in other water units nearby.

One important observation in this regard is that most of the adopter-operators did not use the inputs in accordance with proportion and schedules recommended by the technical experts. They normally chose only those combinations of inputs[1] that suited their resource position best.

It is widely stated that the fish farming involves certain 'cultural practices' to obtain the maximum yield. The tank for fish culture should be neither too deep nor too shallow. If the seasonal tanks dry up during the summer months, lime is required to be applied on the bed. Better results in terms of yield will be achieved if exotic carps like silver carps, grass carps etc. are cultivated along with the indigenous carps like Rahu, Katla and Mrigal. The weed and predatory fishes should be eliminated. Toxicants like mohua oil cake are normally recommended for use because, by its poisoning effect, it eliminates all undesirable elements and simultaneously acts as a manure. Aside from naturally available food chain in the tank, for normal growth, additional food chain in the form of supplementary feeds like rice-bran, wheat-bran, etc. should be given. It follows that intensive composite fish culture involves a high degree of skill.

But it is observed that fish culture is practised in a very neglected and arbitrary manner. Only a small percentage of the adopter-operators in our sample have been found to follow strictly the uses of inputs in accordance with the schedules and doses. In the sample studied, only 7 per cent of the adopter-operators followed preparatory operations like de-weeding, desilting and eradication of predators and weed fishes. Only 4 per cent cultivators used toxicants; 57 per cent of the water had weeds and only 5 per cent of the water units were renovated. The species-mix in seed and the stocking rate—the two major determinants of yield—were not strictly followed. Only 10 per cent of the operators had utilized optimal species-mix in seed and appropriate stocking rate; 28 per cent of the area operated adopted composite fish culture. The operators liberated different species of carps without maintaining right proportions. They could not procure quality fish seed. It was also reported that the average survival from spawn to fry stage is about 30 per cent which is very low. Netting and soil testing were seldom done. It is also to be noted that use of fertilizers and supplementary feed, which indicates the inability of the users to bear higher cash expenses for cultivation and their ability to bear risk, was not made properly. Only 10 per cent of the adopter-operators were found to follow the suggestions made in this regard. Manures for maintaining healthy atmosphere in tank were seldom given. Organic manures were used by about 70 per cent of the operators (the most common being cow dung). Unfortunately, it is also observed that about 95 per cent of the fish farmers in our sample fed their fish irregularly. The most commonly reported fish feed were rice-bran, leftover cooked rice or fresh cow dung, etc. All these findings lend credence to the fact that the level of technology adoption is inadequate, and the input-mix is sub-optimal.

As to why fish farmers are tardy in adopting package of practices becomes clear from the information given in the following Exhibit 6.2. Perhaps the most important reason for non-adoption is the financial difficulties. A sizeable percentage of operators (55 per cent) have reported this problem. Risk factor and failure to provide extension services to provide information also account for the non-adoption. About 64 per cent of the operators consider risk factor as a deterrent for adoption. To sum up, the data reveal that the major causes for non-adoption are the perception of risk factor, cost factor (or financial difficulties) and lack of information.

Risk involved in adopting intensive practices is greater in the sense that it increases proportionately along with an increase in the cost of cultivation. Again, why some operators perceive higher risk in adopting modern practices entailing high costs may be that, in terms of yield, their performance falls much below expectations for various reasons. Consequently, while costs mount up, benefits remain low; the apprehensions about risk and loss thus become real.

Exhibit 6.2: Reasons for Non-adoption of package of practices

Financial Difficulties	*Lack of availability of quality Fish Seed on Time*	*Plurality of Ownership*	*Risk Factor*	*Lack of Proper Extension Services*	*High Price Spread*	*Percentage Reporting Combination*
1	*2*	*3*	*4*	*5*	*6*	*7*
*	*		*			41
*					*	18
	*	*			*	10
*			*	*		8
*	*	*	*			15
60	15	18	74	5	5	100

Note: Figures indicate percentage of farmers reporting this problem.
Source: Field Survey, 2004.

As regards biological and environmental hazards, it is important to note that though initially unwanted predatory fishes or animals can be removed, they may reappear. This poses serious biological hazards. Another serious problem in regard to algal blooms occurring in the ponds is that it creates 'supersaturating' in oxygen levels during day and oxygen 'depletion' during night. This sometimes leads to mass mortality. The 'fear psychosis' of possible occurrence of fish diseases among fish farmers coupled with the lack of fish hospitals is an impediment.

Let us now endeavour to examine the perceptions of the fish

farmers about the factors that hinder higher yield and adoption of modern culture practices. The operators were asked through an open-ended questionnaire to indicate the factors, which, according to them, were responsible for unsatisfactory yield and low adoption and rank the factors according to their degree of importance. The factors ranked 1, 2, 3, 4, 5, 6, 7, 8, 9 and 10 and were given weights 10, 9, 8, 7, 6, 5, 4, 3, 2, 1 respectively. The total rank score for each factor is obtained multiplying the frequency with their respective weightage and adding them up. The factors were then arranged in ascending order of importance on the basis of total rank score and finally ranked. The total rank score has been given in the parentheses against each item in the following Exhibit 6.3.

It is quite evident from Exhibit 6.3 that non-availability of adequate financial assistance is still the most important deterrent factor perceived by a large proportion of the farmers (ranked 1 by all fish farmers). They were concerned with the sufficient amount of credit

Exhibit 6.3: Constraints According to Their Importance

Sl.No.	*Problem areas*	*Rank (Total Score)*
1.	Lack of easily available and sufficient finance in time	1 (358)
2.	Plurality of ownership: absentee ownership	2 (347)
3.	Erratic supply/inadequate availability of quality fish seeds fry/fingerlings in time	5 (287)
4.	High cost of inputs	4 (307)
5.	Adulteration of inputs	6 (275)
6.	Lack of technical knowledge, know-how and training	7 (248)
7.	Lack of contact with extension personnel	9 (198)
8.	Lack of arrangement for soil and water testing at farm level, netting facilities	10 (174)
9.	High price spread	8 (210)
10.	Poaching, theft by unauthorized persons	3 (320)

Source: Compiled from questionnaire.

to be released on time and available easily. The operators also perceived the assured supply of quality fish seeds and fingerlings at reasonable rates at the time of stocking as an important problem (Rank 5). Multiple ownerships of water units also emerged as a problem. The operators ranked it as second. The phenomena of absentee ownership and location of the water units are also considered as a problem. Ponds situated at varying distances between each other and far from the fish farmers' houses may act as a disincentive.[2]

Poaching[3] is also considered as a major inhibiting factor. Poaching or deliberate poisoning due to rivalry, jealousy or enmity is widespread. Fish lifting is taking place particularly in those water-bodies, where there is controversy over ownership rights or situated far from residences. There are also instances that villagers sometimes do not care to understand the proprietary rights or ownership of ponds where they have free access to water units for different purposes. This leads to temptation for poaching where the water units have good stock of quality fish. Other problems which all other operators considered important are lack of training and extension facilities, lack of transport and marketing facilities, non-availability of opportunities for soil and water testing, netting facilities, etc. In short, the fisher farmers recognized both institutional as well as technological or infrastructural problems as constraints.

NOTES

1. Practices considered include de-weeding of ponds, use of fertilizers and fish seed, mohua oil cake, new varieties of seed, extension guidance etc.
2. Ponds are situated at varying distances between each other and far from the cultivator's houses. As the number of ponds in the study area have more than one owner, the sum of distances between them and houses will also tend to increase. It poses the problem for effective supervision and increment of time and distance cost. The farmers also seek to ward off losses and ensure income stability. Under these circumstances, the farmers want to concentrate on those ponds situated nearby. See Michael Chislom's *Rural Settlement and Land Use—An Essay on Location*. Hutchison's University Press, London (1962) Chapter 4.
3. Unlike any other agricultural crops, where the same is ready for harvesting for a shortwhile, fish is ready for harvesting throughout the period of culture.

7

Fish Farmers' Development Agencies

Evaluation

"While the knowledge of torture of someone makes you sick, it is a case of sympathy. While it is not so but you think it is wrong and you are trying to stop it, it's a case of commitment."

— Amartya Sen

"Rational Fools: A Critique on the Foundation of Behavioural Economic Theory" in *Philosophy and Public Affairs,* Vol. 6, Reprinted in *Choice, Welfare and Measurement,* 1982, p. 28.

In recent years, Government of West Bengal has been extending several facilities for expansion of culture fisheries, such as fish seed supply at cheaper rate, subsidy for renovation of tanks as also some welfare measures like Group Personal Accident Insurance, saving-cum-relief scheme, social fishery, river renching etc. Along with these, FFDAs have been established to help the fish farmers with necessary financial, technological and extension support in order to increase fish production in tanks and ponds by composite fish culture practices.[1] The main objectives of the scheme are to provide the much needed field mechanism to popularize composite fish culture among the fish farmers and co-ordinate the activities of different institutionalized agencies in inland fishery development programmes. The focus is on creating a demonstration effect through individuals where success of one fish farmer could motivate others to take up fish farming. This called for :

(a) Progressively bringing all the fallow culture fishing resources under optimum fish production in the area of its operation;

(b) Training and popularizing a new vocation of fish culture;

Exhibit 7.1: Functions of FFDAs

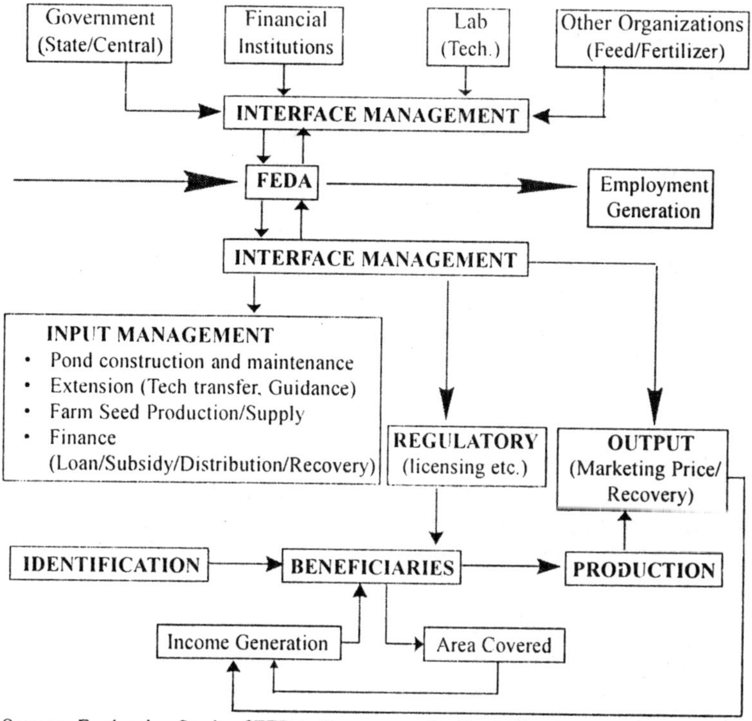

Source: Evaluation Study of FFDAs Programme for Fresh Water Aquaculture, CMA, IIM, Ahmedabad, Dec. 1990, p. 6.

(c) Organizing the FFDAs to serve as a nucleus for spreading fish farming in other areas; and
(d) Strengthening the rural economy by making fish farming economically viable.

Exhibit 7.1 depicts the broad functions of FFDAs.

It was observed that even though traditional fish culture is still practised in different parts of the State, intensive fish culture technology has already started being adopted by quite a large number of common fish farmers in the villages.

Some interesting observations are in order from the data collected from the selected fish farmers during the years 2003 and 2004. In the first place, yields per acre under different types of practices—traditional practices, semi-intensive practices (those who are not FFDA beneficiaries but practise intensive culture) and intensive practices (those who are FFDA beneficiaries and also practise intensive culture) in 2004 are much higher than those in 1989. It is asserted that yield per acre for those farmers who have followed intensive culture practices are much higher than those under semi-intensive or traditional practices in different years. Secondly, yield per acre is gradually increasing at different rates overtime whatever be the type of practices adopted. And, yield per acre under FFDA scheme is much higher than yield under non-FFDA scheme.

The FFDAs deal with a package of approaches to promote intensive fish farming in tanks and ponds. It should be stated that training courses are offered to the identified fish farmers.[2] They are trained on the modern methods of fish culture and fish farm management. Trained farmers' field of information is superior. This superiority of information reduces the risk element in innovation and induces them to adopt new techniques and new inputs. That is, training enhances innovative and allocative efficiency of the farmers to decode information, to evaluate benefit and cost and quickly allocate resources in a cost efficient manner. Again, in close-knit village communities, fish farmers may have interactions with those farmers who have already received training. The emulation of their methods would be an externality of training. It is true that training in fish farming technologies have induced fish farmers and created employment opportunity. Trained persons have been able to create a new cadre of fish farmers. The changes in the yield, if any, in the water units cultivated by the trained farmers are given in the following :

Exhibit 7.2: Practice-wise Distribution of Change in Yield of the Trained Fish Farmers during the years 1989, 1994 and 2004

Year	*Yield Per Acre (in kg)*		
	Traditional Practices	*Semi-intensive Practices*	*Intensive Practices*
1989	458	459	473
1994	582	587	602
2004	589	630	710

Note: The year 1989 is considered, as it was the year of initiation of training for fish farmers under FFDA programme in West Bengal.

Source: Field Survey, 2004.

Yield per acre has significantly increased in the water units of those operators who have received training. The extent to which yields have increased after receiving training may be due, at least in part, to the qualitative improvement in farmer's entrepreneurial ability resulting from the training programme. We have employed the test of association developed by Yule in order to examine whether yield and training programmes are independent attributes. The coefficient of association is found to be positively associated. This exercise leads us to the conclusion that the relationship between training and yield (in other words, use of cultural practices—a proxy for innovations) is quite satisfactory during the period under consideration. Interestingly enough, it is also observed that the idea of adoption of scientific culture has penetrated to and motivated, to some extent, some of those families which did not receive any training whatsoever. It is a reflection of the externality of training.

But it could also be mentioned that the coverage of training is still not very extensive or adequate and training programme has not taken its roots. Some of the trained fish farmers reported some deficiencies in the system. The following Exhibit 7.3 states the responses of the respondents in this regard. It transpires that 63 per cent of the trained farmers reported inadequacy of the training programme because of short duration and non-inclusion of practical aspects and course contents being non-comprehensive. At least 15 per cent of the respondents reported the lack of practical demonstration programme in the system.

Another important component of the programme is the extension programme.[3] The extension programme may serve as a catalytic agent in the diffusion process. Extension agents are supposed to preach the benefits of a new farm technology and this promise is expected to be heard and acted upon by most of the innovative farmers. This is instrumental in creating awareness among the farmers to shift from traditional to modern intensive pisciculture. Motivation of the tradition-bound farmers demands a band of well-groomed committed extension personnel.[4] It should be stated that there are few extension employees who are used to propagate the ideas about technology. The link between the technologies developed in the research institutes and their exposure to the extension personnel were very week. No systems were developed to facilitate constant interaction between the extension personnel and scientists for the continuous feedback. The extension software was also found inadequate. Most of the extension workers had their training just when they entered the services.

Exhibit 7.3: Summary of the Suggestions of the Respondents Regarding Training Programme

Inadequate Time period	*Course content should be revised*	*Practical components like demonstration should be included*	*Percentage Reporting Combination*
1	*2*	*3*	*4*
*	*		63
	*	*	15
		*	0
*		*	7
*	*		0.5
50	48	44	100

Source: Field Survey, 2004.

The operational features of the new technology should be examined against the backdrop of prevailing organizational basis of individual water units and pisciculture as a whole. The new technology is scale-neutral, no doubt; but two problems arise with the farmers in this context – one, as a proposition involving higher cost per acre and second, as a proposition involving more working capital than usual. In this context, there arises the question of financial support necessary to internalize the external economies. The Government of West Bengal through FFDA schemes has made arrangements for giving loans and subsidies to adopt modern intensive fish culture.

Exhibit 7.4 gives the distribution of fish farmers for whom assistance is needed by type of assistance.

From Exhibit 7.4, it is found that finance is the most important source of type of assistance needed in all the years. Technical advice and services are also another important type of assistance, about which 18 per cent of the farmers expressed desire to receive. Another important type of assistance, which the fish farmers desire to obtain, is the regular supply of quantity of inputs and arrangements for different infrastructural facilities.

Exhibit 7.4: Percentage Distribution of the Fish Farmers who needed and Received Assistance

Type of assistance	*Percentage of the Fish Farmers*	
	Needed	*Received*
Finance	95	8
Technical advice/service	90	16
Supply of fish feed and other inputs	90	55
Others (marketing facilities)	100	2

Source: Field Survey, 2004.

But unfortunately, it is revealed from the following Exhibit 7.5 that only a small proportion of the fish farmers have received a considerable assistance. In the case of finance, most of the fish farmers who were desperately in need of finance are, more often that not, deprived. Technical advice and marketing facilities leave much to be desired. Registered and regulated markets are found to be conspicuous by their absence.

As regards the reasons for the unsatisfactory availability of assistances, it is clear from the following Exhibit 7.5, for example, that quite a significant proportion of the fish farmers interviewed were not in the know of the programmes, even though twenty-five per cent of the respondents applied and did not receive financial assistance respectively. Forty-five per cent of the respondents did not even try for technical assistance or extension services, even though fish seed has been received and used by a large proportion of the fish farmers. The rest who did not receive was mainly because of the fact

Exhibit 7.5: Percentage Distribution of the Fish Farmers—Not Receiving Assistance with Reasons Thereof

Reasons	*Finance*	*Technical Advice Extension Services*	*Quality Seed etc.*	*Marketing Facilities*
Did not know	30	30	30	40
Did not try	30	45	50	36
Tried but failed to receive	25	22	10	15
Reluctant to approach because of cumbersome procedure	35	28	20	19

Source: Field Survey, 2004.

that they did not even try (50 per cent of those who did not receive). Marketing facilities through appropriate network have not still developed. Again, cumbersome procedure for applying for receiving assistance also acts as a deterrent factor for the purpose of receiving different assistances from the Government agencies.[5]

There is no denying the fact that there is a clean prospect of FFDAs in facilitating intensive fish culture among the fish farmers and obviously, it created a positive impact. Their role in training, demonstrating the attractiveness of pisciculture and arrangements of loans and seed are found to be instrumental in motivating fish farmers for adoption and hence, increasing the production of fish. The FFDAs are based on decentralized planning ideas. The merits of the FFDAs are derived partly from its ability to discover the dormant resources and skills and endeavour to activate them. Its operative part is being carried out at a very fairly decentralized level – districts, blocks or a cluster of villages. It is true that it is essential to know the fish farmers, understand their social, economic and cultural disabilities and understand their local settings and institutions under which low productivity is generated and sustained before solutions can be conceived of and set into action. The programme is required to be area or community specific, taking into account the different development potentials as well as constraints of each area or community. It is difficult to *tend a flock of sheep from the camel's back*. It is equally difficult to eliminate low productivity from the elevation of the central plan, however 'thorough and earnest'. From these points of view, it is expected that FFDAs would be very much effective. However, it has also been observed that it has left much to be desired. To ensure proper planning, adequate supervision, effective implementation and better monitoring, the FFDAs need to be re-evaluated.

In the first place, it should be stated that this programme, if it is to succeed completely, calls for the real identification of the beneficiaries. It appears from the survey that there exists a great scope for arbitrariness in the selection of beneficiaries. Secondly, the present operational pattern of linking of subsidy with the loan resulted in some unfortunate situation. This has, in fact, resulted in unscrupulous elements taking advantage of the 'subsidy' rather than the real purpose of improving the productivity. Cases of pocketing of subsidy by the bureaucrats, local political leaders and farmers are not infrequent. This needs rigorous monitoring of the schemes at every stage.

Some of the alternative suggestions regarding subsidy may be that an amount may be deposited as fixed deposits in the beneficiaries' account and this could be adjusted after the completion of the repayment by the borrowers. Again, the benefits of subsidy should be linked with the timely repayment. An improvement in the close rapport and co-ordination among the FFDA units, the banks and the local panchayat officers involved in the system is essential.

Thirdly, the cumbersome loaning procedure is normally followed.[6] This causes delays and difficulties in the sanction of loan. The transaction costs of loans, which include informal costs like frequent transportations costs for visiting block and bank officials are quite high. It may be suggested that fishery banks should be established in each block to advance loan well in advance at reasonable rate of interest. It is also observed that the money so borrowed for fish farming has been diverted to other purpose. In this context, it should be noted that if the bottlenecks in the production process dominate, the credit absorptive capacity may go down and the subsidized credit may very well get diverted to consumption purpose. The same consequences may follow here also if the credit does not cover the total requirements of consumption, working capital, etc.

Fourthly, it was also observed that sometimes the flows of credit disbursement are plagued. This is not only due to steadily rising overdue, but also due to, in many cases, faulty loan proposals, laxity in post-credit-follow-up, lack of monitoring of end-use, lending without an integrated programme, etc.

In sum, it should be mentioned that

- credit should be adequate and on reasonable terms;
- there is a need for closer supervision and follow-up to ensure proper end-use of credit;
- real investment should be backed by adequate infrastructural facilities so that potential income can be translated into actual; and
- the assets acquired through credit should generate incremental income (i.e. sufficient to pay back the loan along with interest and still leave reasonable surplus).

Fifthly, FFDAs programmes which have yielded good results in many respects are target-oriented. Being target oriented and in the race to reach the target, benefits are sometimes literally passed on to the ineligibles. It was also observed that the staffs of the development administration were all obsessed with the idea of target achievement. Sometimes, banks are found reactive, instead of proactive. So, it is important to add that target should be fixed, keeping in view the availability of infrastructure and other facilities for optimal utilization of resources.

It is important to note that there has been a proliferation of organizations dealing with development activities like FFDAs, IRDPs, NRY, RLEGP etc. at the grassroots level. This sometimes creates the problem of co-ordination, because they emerge as parallel ladders, each operating in its own style and jurisdiction. This also has resulted in certain negative consequences. That is, rural poor are made to go from pillar to post to have access to the benefits that policy offers to them; and agencies sometimes operate at cross-purposes neutralizing the development efforts. Political interferences also seep into every kind of activities here.

On top of all these, the structure of Panchayati Raj or the Government by the local elective bodies began to totter under the heavy weight of political factionalism, scrambles for control over patronage among politicians and bureaucrats etc.[7] A lot of tension was observed in the system due to conflicts and confrontations between politicians and bureaucrats. Again, in the context of rural power structure and week administrative set up, there is a lot of arbitrariness in the different activities. In an atmosphere of scarcity and unequal distribution of local resources, local officials are found to coalesce with rich cultivators.[8] There are a large number of reported instances of bureaucratic malfeasance. Government officials should also learn how to curry favour with the rural elites. It is not only the rural structure but also the behaviour and attitude of the functionaries that determine the performance and outcome of the different programmes like FFDAs. A certain amount of commitment and enthusiasm is required among the officials engaged in the task.

NOTES

1. The FFDAs are registered societies, envisaging leasing of tanks and ponds owned by government and local bodies to the fish farmers on long term lease, training of farmers, supply of inputs required, provision of financial assistance in the form of subsidy for fish pond improvement and inputs.

The basic organization is the setting up of 58 FFDAs in 58 districts in West Bengal, Bihar, Orissa, M.P., U.P. etc. In addition, extension services were also provided to the fish farmers to spread the technology.

2. Once agents of change appear in the scene in the form of State authorities wanting to change institutions, regulate markets. disseminate research information, it becomes essential for the fish farmers to decode the new information. Ability to decode information depends upon the information field (—know what, why, where, when, how....) and information field can be expanded by

   - 'Learning by doing' method,
   - State effort in terms of extension and demonstration,
   - Information received through social interaction.

   See, for details, D.P. Chaudhuri, *Education, Innovation and Agriculture Development*—Vikash Publishing House, New Delhi, (1979) and K.J. Arrow, *The Economic Implication of Learning by Doing—Review of Economics and Statistics,* Vol. 29 (1962).
3. The training programme is generally followed up with demonstration and extension programmes to combine the fish farmers with the benefit of scientific fish farming. The programme normally includes the principle of fish culture, induced breeding, marketing and management, identification of fish seed, feed and fish disease and its control etc.
4. It has been observed that a large proportion of the rural client, in general, being illiterate, the requirements of complex form filling and documentation have itself limited the access to the bank facilities.
5. A committee headed by the CRAICARD (the Committee to Review the Arrangement of Industrial Credit for Agriculture and Rural Development) chairman Mr. B. Sivaraman has documented the pecuniary and non-pecuniary transaction costs associated with the process of getting a loan. Their conservative estimates put the pecuniary cost between 50 per cent and 75 per cent of the subsidy component, which may account for between 15 and 28 per cent of the total investment.
6. See T.N. Chaturvedi, "Tension in Panchayati Raj, Relations between Officials and Non-officials", *The Economic Weekly,* Bombay, 1964, Vol. XVI, No. 22, pp. 146.
7. R.H. Adam Jr., "Bureaucrats, Peasants and the Dominant Coalition"—*The Journal of Development Studies,* Jan. 1986, pp. 68.
8. The expectation that the new bureaucracy would have new orientation, new activation, new environment and work – way has been belied in actual practice. S. Kothari's and R. Roys, "Relations between politicians and administrators" in *Journal of Public Administration* (IIPA, New Delhi, 1989).

8

Towards Conclusions and Policy Options

It is true that West Bengal has abundant natural endowments, conventional wisdom and human skills in respect of intensified fish farming. But in spite of enormous potentials, much headway has not been made. The levels of production and productivity have not been adequate, leaving a large gap between actual and potential production level. The new technology, which seems technically attractive, has failed to produce substantial results in real world. The recipients of technology are greatly circumscribed by the factors like inadequacy of finance, multiple ownership of tanks and ponds, poaching, deliberate poisoning due to rivalry, enmity or even jealousy, inadequate extension work, existence of strong chain of middlemen, shortage of quality fish seeds and many more others. As a consequence, it can be stated that present unsatisfactory performance can be attributed to all those problems resulting in the absence of rational management and judicious utilization of the resources. Negative socio-economic factors annoying the fish farmers are at work. Let us suggest some measures which may help solve the problem.

Most of the impounded water units (tanks and ponds) do not retain water throughout the year. In the post-winter, depth of water becomes less than the minimum requirement. Farmers are forced to harvest juvenile uneconomic varieties belonging to different size-groups. As a result, they could not fetch a commercially viable market price. Hence, our suggestion is that if the source of water is canal irrigation, water should be released at a very nominal rate. Again, since water is also exhausted due to irrigation or seepage problem, measures may be directed towards rain-fed farming on watershed basis or by strengthening the existing canal supplies. The run-off water collected into farm ponds should be used for irrigation. Ponds should be independent of monsoon for assured water supply. Fallow, derelict water bodies, water logged areas, beds etc. should be developed for culture of air fishes, breathing fishes, catfishes etc. which have high value of consumer demand.

It is also observed that seasonal tanks dry up during summer months. The tank bed should be thoroughly ploughed and exposed to the sun. This will help remove the marsh grasses from the inner depth of the bad area and increase the fertility of the bottom soil. Desilting should be done in those tanks which do not dry up completely. Mohua oil cake should be used to eradicate all predatory factors and other weed fishes from the perennial tanks.

The adoption of the proper culture practices depends on the research and extension efforts of the FFDAs and fish farmers' commitment. Since both these factors varied a great deal across farmers, the farming community, it was likely that the actual culture practices, the intensity of input use and the yield rates did not follow the same pattern. The actual pattern of input use and culture practices adopted by fish farmers and the corresponding production levels attained by them may be understood by focusing on pond construction and improvement, preparation of pond for culture, pattern and intensity of seed stocking, duration of culture, use of feed, use of manure and also the use of labour inputs.

It was observed that vast majority of water bodies are privately owned, particularly in West Bengal. Leasing arrangements are oral and bipartite matter mainly on short-term duration.[1] This is not congenial for good production. Eviction of lessees is quite frequent. Absence of long term leasing policy is a genuine hindrance. The uncertainty attached with short duration normally dampens the fish farmer's motivation. Since there is no guarantee that he would get the same size of pond and under the same conditions as in previous years, the fish farmers are likely to be wary of undertaking investment unless the water units are leased for a long period, say seven or ten years. The financial institutions like banks are sometimes found to be hesitant to lend finance for fish culture if the lease tenure is of short duration. Banks are not certain of the fish farmer's repayment capacity. The pond lease should continue to be considered as collateral for the loans by the banks. Now there should be legal policies to safeguard the interests of the lessees by framing suitable tenancy legislations. There should be long-term lease arrangements. Long-term lease arrangements not only increase production but also provide a full time avocation or employment to the unemployed rural youth.[2]

Both individual holdings and multiple ownerships exist in West Bengal. The State Government should enact legislations to the effect that pond improvement work can be carried on with the consent of one of several co-owners.[3] Some amount of business element should be introduced in the management of multi-owned ponds. If a tank or pond is found, wholly or partly, unutilized for a certain period, the Government may take it over and give it to a cultivator who intends to cultivate it.[4]

Appropriate technology, keeping in view the regional, environmental and ecological differences, should be developed. Induced breeding technology has an edge over other systems of seed production. The technique involves collection of (pituitary hormone) glands from matured carps. But it has been observed that good quality glands are not available. Attempts should be made to build up pituitary banks (with quality control) necessary for breeding both Indian and exotic carps. It should again be noted that some fish farmers had reported the slow growth of fish from the "machine" seed (particularly of Mrigal variety). The temptation on the part of the seed producers to sell fry with its high mortality rates should be avoided. For this, the rearing space should be enhanced. Again, neither the government machinery nor was any standard feed manufacturer in the country made any effort to come up with a standard feed package. Hence, there is a considerable scope for improving the feed by using formulated feeds. Simple procedures like adding fishmeal will improve the conversion factor. The poultry droppings may be used, as is done in some States like Haryana, to act both as feed and fertilizer.

An organized way of exploitation, transportation and marketing should be evolved so that fishes are harvested at regular intervals and earn optimal revenue by sale of fish throughout the year. Bulk landing of fish at a time should also be discouraged. It is also observed that the fish farmers, sometimes in the districts under consideration, for some reason or other, liberate fish fries three or four times in a year and harvest also three or four times from a particular pond. No doubt, this practice would have augmented the supply of fish of size of different sizes, which is sold at Rs. 30.00 to

Rs. 40 per kg. in the local market. This has naturally, on the one hand, increased the per capita consumption of fish of the rural poor people and on the other hand, generated the decline in the supply of table fish.

It is true that in India the three major carps (Katla, Rahu and Mrigal) together with three other exotic carps (Silver Carp, Grass Carp and Common Carp) contribute as much as 90 per cent of the fresh water aquaculture production. It should also be considered important to diversify carp culture with other medium carps (Labio Kalbasu, L. Bata, L. Gonius etc.) due to their consumer preference and reasonably satisfactory a market price in several parts of the country. This calls for greater R&D backup.

Enhancement of production levels from unit area with due emphasis on environmental upkeep and diversification of culture system are the principal approaches to make the aquaculture a sustainable farming practices.

Strengthening of the marketing structures including storage facilities, Ice plants, cold chains, roads and transportations in identified aquaculture areas would ensure higher profit margins to the producers.

Availability of aquaculture inputs to the farmers should receive high priority. Establishment of aqua shop with facilities for soil and water testing the disease diagnosis in addition to availability of material inputs would accelerate the fresh water aquaculture development.

Recent technological advancements have transformed aquaculture to such a status that institutional finance holds the key for the success and growth in this sector.

Mixed farming, i.e. pisciculture in the tank proper and other allied activities such as mulberry plantations, animal husbandry, poultry farming etc. on the embankments of the tanks may be considered for better viability of the Pisciculture scheme. The products of piggery or poultry may be utilized for manuring the water of the tanks. These will provide basic nutrients to water, which also help in maintaining the food chain of the fish. No extra expenditure for manure, supplementary feed and supervision, watch and ward will be required.

Co-operative societies are handicapped by socio-economic limitations.[5, 6] Government efforts in the development of fisheries have been confined to rearing and seed distribution, improvement of tanks and training facilities etc.[7] but no care has been taken to help improve the working of the co-operative societies. As a result, majority of the Primary Fishermen's Co-operative Societies (PFCS) are not working and the existing societies are involved in intra-conflicts, which disturb their smooth functioning. ***More village-based societies should be set up and existing ones are required to be reformed and galvanized.***[8] Again, it is important to note that the cases of inter-linkage of credit and commodity markets in the rural areas which have circumscribed the ability of the government to improve the welfare of the fish farmers should be well taken care of.[9]

There is a need for change of attitude and motivation of the development bureaucrats. It is, of course, rightly pinpointed that the ability to set into motion a process of change would critically depend on the support the planners at the block levels or grassroots levels receive from the political leadership at different levels. To what extent such support would be forthcoming is a moot point. In fact this takes us back to the local socio-political power structure.[10] So, the change in production may be brought about through radical management reform and through changes in the culture of development and bureaucracy.

One point deserves mention. It is that the suggestions or policies imparted to the fish farmers should be economically as well as technically optimal, since most of the fish farmers are in the business of making a living rather than trying to maximize fish production, regardless of the cost. Again, it is worth noting that being poor, the small fish farmers tend to be risk averse. Their predominant goal is economic survival. They really do trade off lower risk against higher profit or the disutility of increasing risk. Adequate stability of output and income and the avoidance of major short-term losses take precedence over profit maximization. They tend to maximize utility rather than income where utility is the function of the variance of income as well as its average level. It should be rightly noted that some ***sort of insurance for fish crop*** might pave the way for stimulating adoption of practices.

In fact, ***monetized development*** does not mean only pumping money into the economy, but it entails that supporting facilities are to be provided for its optimal absorption. This involves also public participation in the measure which is conducive not only to their own ultimate benefits, but to the smooth implementation of the programme. For this, an awareness has to be created among the public and this could be achieved only by closer and direct relationship with the beneficiaries.

Some Emerging Policy Issues

- Fallow, derelict water bodies, water logged areas; beels etc. should be developed for culture. Ponds should be independent of monsoon for assured water supply.
- There is a need for effective rapport of fishery development officials with fish farmers in order to take more responsibilities.[6] Efforts should be made to provide institutional support to train fish farmers in order to enrich and motivate them in the field of scientific fish farming.
- Credit institutions should be galvanized to provide appropriate credit. Credit institutions are required to simplify the procedures for disposal of loan applications.[7] For reducing defaults in repayment of loan, an effective post-disbursement monitoring system should be undertaken.[8]
- Subsidy should be need based. It may be extended to inputs for each year.[9]
- A system should have to be organized to facilitate constant interaction between the extension personnel and the scientists for feedback. The extension software should be carefully developed.
- The trainers in the training institutions should also be periodically exposed to new knowledge and skills. The training they impart is then expected to be meaningful.
- It would be appropriate to provide facilities for the supply of fingerlings and evolve locally suitable combinations.
- Steps should be taken to lease out ponds for a longer period, that is, at least ten years or more, to ensure fish pond development activity a financially viable proposition.
- Steps should be taken to help fish farmers in setting up marketing co-operatives with vertical linkage at the terminal and transit markets.

- Aquaculture needs to be integrated with agriculture, piggery, duckery etc.
- Aquaculture needs to be treated on par with agriculture for tax, water supply, electricity and other facilities.[10, 11]

It is true that liberalization of tariff and quotas in the fisheries sector may encourage more trade and help to derive more benefit from increased access to institutional market. At the same time, there is a growing concern that trade liberalization through its different provision like use of regulatory barriers such found safety regulation, quality and composition standards, labelling requirements, technical barriers to trade (TBT) and Sanitary and Photo Sanitary (SPS) measures might have some adverse effects on the exports of fish from the countries like India. In the event of the removal of quantitative restrictions on 60 items of fish and fish products in the year 2001 fishermen are apprehending the cash in price because of large-scale import of fish. Provisions like Trade Related and Intellectual Property Rights (TRIPR) and Trade Related Investment Measures are likely to have some impact upon fishing industry, the conservation of fisheries resources and the communities depending upon them. In the light of the different provisions like sanitary and phyto-sanitary measures (SPS) and Technical Barriers to Trade (TBT), countries like India intending to develop fisheries sector for exporting fish and fish products to different countries are required to follow the Hazard Analysis Critical Control Points (HACCP). To meet the challenges of different threats under SPS and TBT, Indian processing industries are to take care of different steps for improving parameters consistent with international standards.

In order to protect our fisheries from increase in imports and at the same time to improve competitiveness in the global market, there is a need for reduction in production costs, increase in productivity and also improvement in the quality of fish products.

The suggestion made here are more general and good as far as it goes. But it may not go really very far. Pisciculture is a case of 'diversity in unity'. So, it calls for detailed individual farm plans in each case.

Fig. 8.1: Various Aspects of Inland Fish Production and Training

NOTES

1. The Commission has suggested a lease period of fifteen years of ponds with further extension in case of satisfactory progress. Report of the National Commission of Agriculture, Vol. VIII, P-66, Ministry of Agriculture and Irrigation, Government of India (1976).
2. According to the West Bengal Land Reform Act, 1965 (amended by an enactment in 1981), all classes of "land" including tanks are to be covered within the purview of ceilings. As a consequence, ponds and tanks in excess of ceiling and its distribution have financial, administrative and legal complications. – S. Paul's "Infrastructural Base Support Measures for Aquaculture: A Case Study of West Bengal" in *Fisheries Development in India* by U.K. Srivastava & Reddy (ed), Concept Publishing House, New Delhi.
3. It is heartening to note that the West Bengal Inland Fisheries Act, 1984 seeks to tackle the problem of multi-ownership by empowering the State Government to take over derelict ponds and lease them out to a group of fish farmers or co-operatives of genuine farmers.
4. It has been pointed out that the tanks should be leased to a person/persons among the shareholders who will guarantee the production of the targeted optimum; other shareholders will get their shares according to the target. Excess production will be accounted as profits by the person/persons who will take the responsibility for its management as entrepreneurial loss or gain, if any, will be borne by them as a guarantee against negligence. – See Susil Kumar Chakraborty's "Some Problems of Fish Culture in West Bengal: A Case Study", *Indian Journal of Agricultural Economics*, Oct.–Dec. 1968, Vol. XXIII, No. 4, p. 230.

5. In this connection, it may be noted that the private owners of water resources (particularly small) cannot invest in ponds development. They are lacking in resources and they do not have access to the organized money market. It has been observed in the study area that some fish cultivators in the districts are compelled to borrow for meeting consumption requirements and working capital for fish farming. They are found indebted to private financiers like commission agents aratdars etc. Again, to secure their access to the fish landings, the commission agents or aratdars advance loans against the stipulation that the catch should be disposed of through them. These agents act simultaneously as a lender and auctioneers. The financial dependency of some fish producers on the middlemen sometime robs them of their initiatives, investment and planning. This interpenetrations of the credit sphere with the production marketing sphere illustrates the case of interlinkage of product and credit markets.
6. The National Council of Applied Economic Research and World Bank both have found the schemes as the most suitable organizational framework and as a nucleus of growth.
7. It has been observed that a large proportion of rural client, in general, being illiterate, the requirement of complex form filling and documentation have itself limited the access to the bank facilities.
8. Some scholars, taking a cynical view, have termed bribes a useful lubricant for cumbersome administrative machinery. (See "Corruption and Political Development – A Cost Benefit Analysis", J.S. Nye in *American Political Science Review*, June 1967, p. 417). It is also stated that corruption in LDCs stems from the power to perform or not to perform acts that results in economic benefits to individuals (in "*Corruption and Development*" by Been Stock, M. in World Development, Vol. 7, January 1979, p. 16.).
9. Mr. A.K. Ghorai has suggested the introduction of different subsidy or privileges for different purposes to different beneficiaries. He also mentioned that in Japan, State subsidies are generally granted only for common facilities. See A.K. Ghorai's article "*Modernization of Small Scale Fish Seed Units, Quality Control and Marketing*", presented in the Second National Fish Seed Congress held at Calcutta, 1986.
10. A committee headed by the CRAFICARD (the Committee to Review the Arrangement of Industrial Credit for Agriculture and Rural Development) chairman Mr. B. Sivaraman has documented the pecuniary and non-pecuniary transaction costs associated with the process of getting a loan. Their conservative estimates put the pecuniary cost between 50 per cent and 75 per cent of the subsidy component, which may account for between 15 and 28 per cent of the total investment.
11. Report by the working group on block level planning (chairman – Professor M.L. Dantwalla), Planning Commission, Government of India.

Questionnaire

STUDY ON "AN ENQUIRY INTO THE NATURE AND CONSTRAINTS IN THE DEVELOPMENT OF INLAND FISH CULTIVATION"
(SPONSORED BY UNIVERSITY GRANTS COMMISSION)

1. Identification:

1. Name of the cultivator ....
2. Religion .... Caste ....
3. Village .... Mouza ....
4. Anchal Panchayat ....
5. Block .... District ....
6. Reference Year ....
7. Name of the Investigator ....
8. Date of Investigation ....

2. Total Members of the Family:

1. Adult : Male .... Female ....
2. Minor : Male .... Female ....

3. Details of Working Members:

Sl. No of Workers	*Sex*	*Age*	*Education*		*Occupation*
			Primary	*Secondary*	

4. Type/Size of the farm (Agriculture):

Types of Land	*Size (Acre)*	*Irrigated*	*Non-Irrigated*
1. Owned land 2. Leased in Land 3. Leased out Land 4. Operated Land 5. Current Fallow 6. Vested Land 7. Any Other			

5. Size of the water area of the ponds/tanks (in Acre):

Types of the Ponds	*2001-2002*	*2002-2003*	*2003-2004*
1. Owned Pond 2. Leased out Pond 3. Leased in Pond 4. Operated Water Area 5. Current Unused or Fallow			

6. Material inputs cost (Rs./kg):

Item	*2001-2002*	*2002-2003*	*2003-2004*
1. Fish seeds 2. Manure 3. Fertilizer 4. Pesticides 5. Labour cost 6. Fixed cost			

7. Disposal of Fish Production:

Channels of Disposal	*How disposed*	*2001-2002*	*2002-2003*	*2003-2004*
1. Wholly for personal consumption 2. Mainly personal consumption and partly marketing 3. Mainly marketing 4. Mainly marketing and partly personal consumption				

8. Prices of the Fish:

Name of the Fish Varieties	*Price at which fish is sold*			*Market Price of fish*		
	2004-03	*2003-02*	*2002-01*	*2004-03*	*2003-02*	*2002-01*

9. Are you satisfied with price?

10. If not, what price do you expect?

11. What kind of constraints do you face in this case?

12. Wholesale price and Retail price:

Varieties of Fish	*2004-2003*		*2003-2002*		*2002-2001*	
	Wholesale	*Retail*	*Wholesale*	*Retail*	*Wholesale*	*Retail*

13. Income of informants by sources over years:

Sources	*2004-2003*	*2003-2002*	*2002-2001*
Agriculture			
Fishery			
Horticultural Crop			
Forestry			
Arts and Crafts			
Trade			
Transport			
Services			
Profession			
Agriculture Wages			
Non-agriculture Wages			
Remittance received/sent			

14. Package of Practices Adopted in Pisciculture:

Items	*Whether used*	*Source*	*2004-03*		*2003-2002*		*2002-2001*	
			Qty.	*Value*	*Qty.*	*Value*	*Qty.*	*Value*
Seeds								
Manure								
Fertilizer								
Pesticides								
Fish meal								
Feeds								

15. Borrowings:

Source of Borrowings	*Year*	*Amount*	*Rate of Interest*	*Purpose*	*Amount of default*	*Reason/s for Default*

16. Various Problems Regarding Adoption of Modern Culture Practices:

Various Problems	*Whether faced or not by informant*			*Ranking of problem*
	2004-03	*2003-02*	*2002-01*	
1. Problem of lack of knowledge				
2. Non-planning				
3. Erratic supply of seeds and inadequate availability				
4. Problem of marketing				
5. Maintenance of ponds				
6. Problem of netting				
7. Problem of catching				
8. Problem of distance (location)				
9. Problem of finance				
10. Problem of ownership				
11. Risk factor and failure of extension services				
12. Lack of technical knowledge, training and supervision				
13. Problem of poaching				
14. Problem of mortality				
15. Problem of stealing				

17. Some other Investigations made:

State the production level in different years (kg/acre)

Ponds/tanks	*2000*	*2001*	*2002*	*2003*	*2004*
Owned pond/tanks Leased in pond/tanks					

1. Between Agriculture and Pisciculture to which do you accord greater priority and why? Give reasons for your choice.
2. Do you consider the use of water for Rabi crop cultivation at the cost of pisciculture? Is it a more remunerative proposition? Yes/No
3. If yes, why? ....
4. Is distance of location of water area a constraint? Yes/No
5. If yes, why? ....
6. Is the pond owned by you and cultivated by you? Yes/No
7. Do you consider absentee ownership a constraint? Yes/No
8. If yes, how? ....
9. In case of multi-ownership, how many shareholders are there?
10. How did you get water unit (Pond or tank): Inherit /Acquired through price given by you?
11. Do you wholly own the water unit (pond or tank): Yes / No
12. If not, what is your share? ....
13. Do you consider multiple ownership a constraint? Yes/No
14. If yes, State the ways in which it acts as a deterrent. ....
15. How do you consider a leased in pond less productive/more productive/equal productive in comparison to owned pond? State production per acre in your pond: ....
16. Do you use modern varieties of fish seed? Yes/No
17. If yes, according to schedule or according to resource availability?
18. If not, what factors do you consider important (arrange according to priority) in not using varieties of seeds? High Price/Non-availability/Lack of interest/Lack of knowledge
19. Do you sell your produce in the market? Yes/No
20. If yes, what is the channel?
21. Do you consider the price you receive appropriate? Yes/No

22. Do you have any idea about the final consumer's price? Yes/No
23. What suggestion do you like to make in order to improve marketing efficiency?
24. Are you aware about the services available with regard to technical knowledge about pisciculture? Yes/No
25. Do you face any problem regarding shortage of water?
26. What are the major problems for you in scientific (composite) fish farming?
27. What factors deter you from adopting scientific (composite) fish farming?
28. Are you aware of the recommended schedule of stocking of fish seed? What is your source of collection of fish seed?
    ....
29. Do you recognize the length of duration of lease term as factor for adopting intensive fish cultivation?
30. If yes, how are you benefited out of it?
31. Do you get any assistance from FFDA schemes?
32. Do you consider finance as a constraint?
33. In case of financial difficulties, how do you solve problem?
34. Do you consider poaching/stealing a constraint?
35. What suggestion do you like to make in this regard?
    ....
    ....
36. Do you encounter any problem like fish disease?
    ....
    ....
37. If yes, what preventive measures do you normally adopt?
    ....
    ....
    ....
38. What kind of suggestions do you like to make to improve the working of the FFDA and other Government Programme?
    ....
    ....
    ....

Date: Signature of the Investigator

Glossary

Algae: Members of a group of mostly aquatic, photosynthetic organisms that defy precise definition. Algae provide much of earth's oxygen, serve as the food base for almost all-aquatic life and provide foods and industrial products. Algae are common as green sheens on pools and ponds. Use of algae is perhaps as old as mankind. Many species are eaten by coastal societies.

Aquatic Ecosystem: Any system of interacting aquatic life within its natural and physical environment.

Aquatic Environment: It means the natural and biological resources comprising any aquatic ecosystem.

Brackish Water: Water having a mineral content intermediate between fresh water and sea water (1000 – 10,000 mg/*l* dissolved solids).

Co-management: A valuable tool for the sustainable management in fisheries. It can also help in resolution of potential conflicts among user groups. As part of co-management, effective dialogue between local and tourism communities is needed, as it is a commitment on the part of non-resident stakeholders.

Dayabhaga: One of Hindu laws of Inheritances. The *Dayabhaga* system provides that sons and grandsons had right in the family property even during the lifetime of the father.

Ecosystem: An integrated unit consisting of the community of living organism and the non-living, or physical, environment in a particular area such as a tropical rainforest, coral reef or grassland. The relationship among species in an ecosystem is usually complex and finally balanced, and removal of any one species can often be disastrous.

Enhancements: These are interventions in the life/rearing cycle of the common pool resources. Enhancement technologies involve culture based fisheries/ranching, habitat modification, fertilizations, feeding and eliminations of predators. They use little feed and energy and provide high returns and inputs of labour and capitals. Resource poor sections of the population benefit greatly from the enhancements, which help in generating food, income and other benefits from under-utilized, new or presently degraded aquatic resources.

Fishery: One or more stocks or species of fish or aquatic life (usually fresh water) that can be treated as a unit for conservation/ management purposes.

Food Chain: A sequence of plants and/or animals generally increasing in complexity and decreasing in numbers with each succeeding species feeding upon its predecessors.

Freshwater: Water having a low mineral content acceptable for potable purposes (< 1000 mg/*l* dissolved solids).

Freshwater Ecosystem: Include rivers, lakes, freshwater wetlands, geothermal ecosystems and underground aquifers.

Fish Farmers' Development Agencies: The FFDAs are registered societies, envisaging leasing of tanks and ponds owned by government and local bodies to the fish farmers on long term lease, training of farmers, supply of inputs required, and provision of financial assistance in the form of subsidy for fish pond improvement and inputs. In addition, the extension services were provided to the fish farmers to spread the technology.

Glass Jar Hatchery: It is an example of low cost Chinese type portable hatchery, especially in respect of fresh water variety of fish species, which Government of India has helped to promote in rural and semi-urban areas.

Habitat: The place or type of site where a biological community naturally occurs.

Hindu Law of Inheritance: *Mitakshara* and *Dayabhaga* were two laws in ancient India, the dominance of which continued uninterrupted in the countryside has been playing or played an important role in the subdivision and fragmentation of tanks and ponds in India.

Inland Fisheries: Any activities conducted to extract fish and other aquatic organism from inland waters.

Jhora Fishery: In the hill areas of Darjeeling, FFDAs have taken up pisciculture by impounding running water from hilly streams, which is known as 'Jhora'.

Mangrove Swamp: Muddy swamp found on tropical coasts and estuaries, characterized by dense thickets of mangrove trees that are adapted to live in creeks of salt water and send down special breathing roots from their branches to take in oxygen from the air.

Mitaksara: One of Hindu laws of Inheritances. According to the law, father was the steward of the property. He could dispose of the property and if he died, property succeeded to his sons who would divide them equally. The eldest son, having no special claim, could get a little consideration amounting to twentieth part of the share.

Nursery pond: This type of pond is used for nursing 2.5 to 7.5 cm fry until the desired size of fish is reached.

Reservoir: A natural or artificial, usually open, impoundment or body of water to be used for recreation, power, flood control or drinking purposes is stored.

Sewage Fed Fishery: A unique example of recycling urban waste for productive use in pisciculture. Sewage Fed Fishery in the eastern fringe of Calcutta is an eye-opener to the world and show how the urban waste can be purified through pisciculture. It not only maintains the eco-system but also continues a major source of fish supply.

Social Fishery: This scheme was introduced in 1987-88 by the Fisheries Department aiming at stocking water bodies owned by the Government, institutions etc. by transferring management to group of persons selected by Panchayats, from among those, living below the poverty line.

Stocking: The protection of natural fish resources is a cost effective strategy for increasing fish production in inland waters but the effect is sometimes limited by environmental conditions. In those cases, where the composition of natural fish species is not optimal for natural productivity, stocking can effectively increase fish production in inland waters. It involves the release of artificially produced fish seeds into natural water bodies and usually depends on the artificial breeding of the fish species. Two major objectives of stocking are to enhance fish resources, and to establish extensive fish culture production.

Sustainable Development (SD): SD is a requirement to the present generation to manage the resource base such that average quality of life we ensure ourselves should potentially be shared by other future generations.

Sustainable Yield Cropping: The removal of surplus individuals from a population so that the population maintains a constant size. This usually requires selective removals of animals of all ages and both sexes to ensure a balance population structure. Taking too many individuals can result in a population decline, as in over fishing.

System Approach to Aquaculture: This multidisciplinary approach involves the following steps – stating the problem, identifying the system, classifying and describing the system boundaries and critical factors, analyzing the problem, proposing solution to the problem based on system understanding, testing the solution, and implementing the knowledge on how to solve the problem.

Threatened Species: A species or community that is vulnerable, endangered or presumed extinct.

Wetland: Permanently or intermittently wet areas, sallow water and land water margins that support a natural ecosystem of plants and animals that are adapted to wet conditions.

Zooplankton: Small animals or the larvae of larger animals, which drift in the surface waters of the sea and lakes.

ANNEXURES

Annexure 1

Relationship between Stocking of Inputs—Fish Seeds and Productivity—A Note

It is pertinent to note that the stocking rate and the species mix in fish seeds are considered two important determinants for production of fish. Quite a significant proportion of fish farmers used to practise composite fish culture in our study area. This includes major carps, exotic carps, minor carps etc. Fish seed stocked in a sample ponds may be categorized into two groups: exotic carps and indigenous carps. The exotic carps include common carp, grass carp and silver carp; whereas, the indigenous carps include major carp, minor carp etc.

Exhibit A : Size-classwise distribution of quantity and quality of fish seed stocked (per acre) in percentage

Species	*Small*	*Medium*	*Large*	*All ponds*
Indigenous	70	58	72	76 (12 kg)
Exotic	30	42	28	24 (4 kg)
Total	16 kg	18 kg	15 kg	16 kg

Source: Field Survey, 2004.

Per hectare stocking rate of fish seeds differs among the fish farmers. It is clear from the Exhibit A that, on an average per acre 15 thousand of fish seeds (amounting to 16 kg) were stocked. 76 per cent of them constituted indigenous and 24 per cent exotic fish species. Size-classwise variations in the stocking rate of fish seed were also observed. Stocking rate was found highest in the medium sized ponds (18 kg per acre) followed by small ponds (16 kg per acre) and large ponds (15 kg per acre). Small ponds owners used to stock poor quality of fish seeds which were generally collected from natural sources which had high mortality. Large sized pond owners usually used comparatively large sized fish seeds than small and medium pond owners. The medium pond owners stocked fishponds by comparatively larger proportion of exotic species of fish seeds (42 per cent) than the corresponding proportion in the small ponds (30 per cent) and large ponds owners (28 per cent).

Attempt was also made to examine the proportion of the different species of fish *vis-à-vis* their recommended proportions for stocking in multi-carp fish culture. There are four major groups of carps like upper level carps (Catla and Silver), middle level (Rohu), lower level (Mrigal and common) and grass eating carps (Grass Carps). In the following Exhibit B, the proportion of different groups of fish seed stocked in three separate categories of ponds are presented.

Exhibit B

Groups of Species	*Recommended Proportions*	*Small*	*Medium*	*Large*
Surface Feeder	30-40	30	29	21
Column Feeder	10-20	30	18	17
Bottom Feeder	40-50	35	35	40
Grass Eating Feeder	5-15	5	18	22

Source: Field Survey, 2004.

The stocking rate, as it is evident, was not unsatisfactory; but, in most cases, it was found not according to the recommended schedule. The composition of different species of fish seeds in multi-carp culture was mis-matched to the recommendation. A large number of sample farmers expressed their completely unawareness about the proper stocking proportion. Stocking of small size of fish seeds, which have high mortality rate, was found predominant. Also, 50 per cent of the fish farmers could get indigenous and exotic carps on ponds, since the traders used to deliver fish seeds on fishpond sites. Even though the fish farmers used to collect fish seeds from different sources (like fish farmers, traders, Fishermens' Co-operative Societies), the fish traders emerge as a most important source, since in our sample about 69 per cent of the fish farmers collected fish seeds (indigenous major carps, exotic carps) from them.

Annexure 2

Culture Practices

Fish cultural practices include all activities from pre-stocking operations to harvesting of fish through human efforts. The cultural practices may simply be categorized on the basis of major operations like pre-stocking, stocking, input application, applications of supplementary feeds and harvesting.

Pre-stocking refers to those operations prior to stocking in order to clean the pond water by deweeding, netting or other manual activities to eradicate plan weeds, insects, frogs, predatory fish which adversely affect the level of fish production. This would facilitate the better environment for proper growth and movement of fingerlings. **Stocking** refers to applications of external fish seeds into the pond for the purpose of raising them up to marketable size. There are two important types of fish culture prevailing in the study area which are known as indigenous major carp culture and exotic carp culture. Indigenous major carp culture includes fish species like Rahu, Catla, Mrigal etc. whereas, exotic carp culture include the culture of exotic species, like Silver Carp, Common Carp and Grass Carp.

Fish thrives well in alkaline water. Lime is used in a pond to maintain the appropriate alkaline nature of the fishponds. **Toxicants** are the substances which are used to eradicate predatory fish, insects, and frogs from the ponds. The most common toxicant is Mahua Oil Cake. Manures are not directly consumed by the fish. It, in fact, enhances the vegetative growth of green aquatic plants and phytoplankton which are ultimately consumed by the phytophagous fish. Dung is one of the most commonly used manure. The purpose of adding **fertilizer** in the pond is almost similar to that of manure. It also facilitates the growth of micro and macro-vegetative plants for an ultimate consumption by the phytophagous fish. The most common fertilizers used are urea and single super phosphate. Farmers very often expressed their unawareness in this matter. All these may be largely due to the inadequate institutional efforts to acquaint them regarding the use and benefits of manure and fertilizers in fish production. Another critical inputs necessary for better and vast growth of fish is the use of *supplementary feed* like rice bran, mustard oil cake etc. In almost all the ponds, upon enquiry, it was revealed that very few farmers use supplementary feed as per recommended schedule, rather hit and trial method is followed.

Annexure 3

Sources of Collection of Fish Seeds

While conducting survey, it was also revealed that fish farmers used to change very often the sources of collection of their fish seeds. The reasons identified for this change were higher prices, poor quality, inadequate quantity, distance places, uncertainty about quality, delay in availability, transportation problems etc. poor quality of the fish seeds was the most prominent reason (Exhibit C).

Exhibit C : Reasons for Changes In the Sources for Fish Seeds in Order

Reasons	*Rank*
Poor Quality	1
Higher Prices	2
Distant Place	3
Delay and Uncertainty Regarding Availability	4
Transportation Problems	5

Note: Poor quality of fish seeds may be understood in terms of improper size, higher mortality rate, mixed species etc.

Source: Field Survey, 2004.

Small fishpond owners tend to stock much higher quantity and large number of fingerlings. It was mainly due to use of locally collected fingerlings from rivers. Poor fish farmers neither had easy access to fish seeds from nursery located at different places nor had funds to purchase better quality of fish seeds.

Annexure 4

Conversion Ratio of Fish from Fish Seeds

Fish seed is one of the most indispensable input for the production of fish. In order to obtain the higher level of fish production, there is a need of use of appropriate number and quantity of quality fish seeds. To understand the production efficiency in fish production, conversion ratio is an important measure. Use of higher potential species of fish would yield higher quantity of fish.

In our study, on an average, the conversion ratio was found to be 64.5, indicating that the use of one kg of fish seeds would produce 64.5 kg of matured fish. It has also been observed that, on an average, exotic fish species have comparatively higher conversion ratio (71) than that of indigenous fish species (65). Again, silver carp, one of the exotic fish, have higher conversion ratio (89) than grass carp (81) and common carp (75). On the other hand, Rohu, one of the indigenous major carps, had comparatively higher conversion ratio than Mrigal (66) and Catla (61) (Exhibit D).

It was also observed that the medium ponds had much higher conversion ratio for all the species under investigation. This was followed by small ponds and large ponds. The conversion ratio of silver carp of exotic species was found to be higher in all the categories of ponds. This was found true also in the case of Rohu of indigenous carp on all categories of ponds. In point of fact, it may be stated that the fish farmers are expected to use higher proportion of fish seed species which would have higher conversion ratio. In our study area, relatively a smaller proportion of exotic species of fish seeds like silver carp which used to have higher conversion ratio was used by the farmers of all categories even though larger proportion of Rohu in stocking was used. Upon enquiry, it was also observed that the demand for silver carp may be low due to poor test of this species;

hence, the fish farmers are reluctant this variety. The conversion ratio may not necessarily be a major determinant in the choice of fish species for stocking. Factors like market price, demand and test of a particular variety of seeds may be some other relevant determinants.

Exhibit D : Conversion Ratio (Fish Species-wise)

Species	*Average*
Rohu	69
Catla	61
Mrigal	66
Indigenous fish	65
Silver carp	89
Grass carp	81
Common carp	75
Total Exotic Fish	71
All Types of Fish	64.5

Source: Field Survey, 2004.

Annexure 4A

Inland sector remains a sector of much promise. Unlike agriculture, because of expansion of culture based fisheries enterprises, the contribution of inland fisheries sector to grow domestic product has been increasing at a significant rate. This sector is complex enterprises which operates under the intensified network of natural resources, other enterprises with fisheries and other socio-political variables. This is evident from the following Exhibit E that during the period1995-96 to 2001-03 the percentage of possible changes in output values with respect to time has increased by 30 per cent with inland fisheries, constituting more than 55 per cent of the growth. The rate of growth of value of output of the fisheries sector has been targeted 7 per cent per annum during the fourth and element during five year plan (GOI, 1999). Fish consumption is also expected to increase from 5.05 kg/capita in 1996-97 to 9.30 kg/capita in 2011-12 (GOI, 1999) (Exhibit E).

Exhibit E : Change in Value of Output in Fisheries, Agriculture etc.

Sector	*1995-96*	*1996-97*	*1997-98*	*1998-99*	*1999-2000*	*2000-01*	*2001-02*
1	*2*	*3*	*4*	*5*	*6*	*7*	*8*
Agriculture	103.60	113.65	110.19	118.68	119.43	116.43	122.40
Fisheries	112.25	117.77	120.75	128.20	139.70	140.78	145.78

Note: This is of course worked out on the basis of expected population growth rate, and per capita consumption expenditure and the elasticities of demand with respect to per capita consumption expenditure.

Ref: Government of India (1999), Ninth Five Year Plan, 1997-2002, Vol. 1, Development Goals – Strategy and Policies, Planning Commission, New Delhi.

Source: National Accounts Statistics, Central Statistical Organization.

Annexure 5

Inland Fish Production by States/Union Territories, 1990-2000 ('000 Tonnes)

Sl. No.	*State/Union Territory*	*1990-91*	*92-93*	*94-95*	*95-96*	*96-97*	*97-98*	*98-99*	*99-00*
1.	Andhra Pradesh	136.25	151.48	195.13	203.97	207.31	226.31	260.83	380.58
2.	Arunachal Pradesh	1.25	1.60	1.80	1.85	1.85	2.313	2.30	2.40
3.	Assam	76.00	140.00	153.00	155.06	154.61	155.13	155.71	159.77
4	Bihar	159.93	164.07	195.37	239.58	249.78	208.54	202.29	254.74
5.	Goa	3.05	2.77	3.44	3.71	3.30	3.24	3.18	3.51
6.	Gujarat	45.00	49.00	70.10	60.00	65.28	70.80	80.00	70.33
7.	Haryana	23.20	20.15	24.13	28.01	30.00	32.05	32.52	30.00
8.	Himachal Pradesh	5.20	6.39	5.29	5.94	6.27	6.69	6.79	7.00
9.	Jammu & Kashmir	13.00	14.30	16.10	16.52	17.80	18.53	18.85	19.01
10.	Karnataka	53.00	65.70	70.29	87.35	101.65	95.28	95.00	126.65
11.	Kerala	36.34	38.16	48.19	49.59	52.10	57.51	65.85	73.90
12.	Madhya Pradesh	36.95	38.16	48.19	49.59	52.10	57.15	65.56	73.90
13.	Maharastra	64.00	77.19	89.99	77.00	109.00	127.00	125.50	135.39
14.	Manipur	8.50	11.20	12.01	12.50	12.71	13.70	15.39	15.51
15.	Meghalaya	1.52	3.56	3.95	3.58	3.58	3.09	4.55	5.00
16.	Mizoram	2.95	3.38	2.00	2.50	2.55	2.70	2.75	2.89
17.	Nagaland	0.83	1.50	2.50	3.00	4.00	3.70	4.54	5.00
18.	Orissa	836.29	93.76	134.77	134.85	134.50	153.43	150.90	135.30
19.	Punjab	11.20	16.00	24.00	26.00	32.00	36.00	44.50	47.18
20.	Rajasthan	6.02	10.92	13.97	12.40	14.30	15.10	12.00	12.97
21.	Sikkim	0.10	0.09	0.10	0.15	0.15	0.14	0.14	0.14
22.	Tamil Nadu	82.00	98.00	108.00	108.00	109.00	109.50	110.20	112.00
23.	Tripura	21.20	23.20	25.10	25.71	27.47	27.91	28.30	29.34
24.	Uttar Pradesh	104.26	121.43	139.90	145.70	149.43	160.02	183.03	192.71
25.	West Bengal	555.00	612.00	669.22	740.00	765.00	786.02	823.50	865.70
26.	Andaman & Nicobar	0.00	0.10	0.05	0.05	0.04	0.04	0.05	0.06
27.	Chandigarh	0.05	0.05	0.08	0.08	0.08	0.00	0.00	0.00
28.	Delhi	3.00	3.60	3.90	4.00	4.10	4.20	4.42	4.30
29.	Pondicherry	3.16	3.55	4.19	4.00	0.04	4.10	4.11	4.21
	INDIA	1536.25	1789.0	3096.7	2242.3	2381.4	2438.0	2568.81	2822.7

Annexure 6

Statistical Profile of Fishery Development in West Bengal

General Information

Geographical Situation	:	Latitude 21°38′ North to Latitude 27°10′ North, Longitude 85°38′ East to Longitude 90°50′ east
Area	:	88,551 sq.km
Population	:	(2001 Census) 8.02 cr.
Fishermen village	:	11,200
Fisher Folk	:	13,87,883
Marine village	:	652
Inland Fishermen	:	12,88,027
Marine Fishermen	:	1,99,122
Coast Line	:	158 km (excluding Hooghly-Matlah Estuary)
Average Rainfall	:	120cm to 400cm (175cm)
Fishery Resources	:	

A. Inland Sector

Sl. No.	*Type of Fishery*	*Total Potential Resources (in Lakh Hectare)*	*Under culture (In Lakh) Hectare)*	*Percentage of Resource Area under Culture*
1.	Ponds/Tanks	2.76	2.20	79.71
2.	Beel & Boor	0.41	0.21	51.21
3.	Reservoir	0.16	0.03	18.75
4.	(a) River	1.72	—	—
	(b) Canal	0.80	—	—
5.	Sewage-fed Fishery	0.04	0.04	100.00
6.	Brackish Water Fishery	2.10	0.48	22.85

B. Impounded Water Area (Tanks/Ponds)

Sl.No.	*Item*	*Area (In hectare)*	*Percentage over total*
1.	Culturable	1,94,113.88	70.28
2.	Semi-derelict	55,162.96	19.97
3.	Derelict	26,925.00	9.75
4.	Total	2,76,201.84	100.00

C. Marine Sector

Sl.No.	*Item*	*Area (In hectare)*	*Percentage over total*
1.	Inshore	Up to 10	770
2.	Offshore	10 – 40	1813
3.	Continental Shelf	Up to 100	17049

1. Estimated demand of Fish: 11.5 lakh tonnes (at the end of Ninth Plan)
2. Number of minor fishing harbour: 2
3. Number of Ice Plants: 130
4. Number of Processing Plants: 32
   Processing Capacity : 269 MT/day
5. Total number of Cold Storages: 36
   Capacity: 2468 MT
6. Temperature and Salinity:
   Sea Surface Temperature: 20°C
   Salinity : Above 34 ppt.
   Brackish Water : Up to 20 ppt.
7. Fishery Districts : 18
8. Fishery Blocks : 306
9. Primary Fishermen's Co-operative Socieities : 1128
   (i) Inland : 967
   (ii) Brackish Water : 30
   (iii) Marine : 131
10. Central Fishermen's Co-operative Societies: 20
11. State Level Fishermen's Co-operative Society: 1

Annexure 7

District-wise Impounded Water Area (Ponds & Tanks) in West Bengal (in hectare)

Sl. No.	*Name of the District*	*Culturable Area*	*Semi-Derelict Area*	*Derelict Area*	*Total Area*
1.	Coochbehar	1598.11	1993.76	727.44	4319.31
2.	Jalpaiguri	386.36	468.00	101.41	955.77
3.	Uttar & Dakshin Dinajpur	9678.66	2173.36	2281.76	14133.78
4.	Malda	4579.56	900.13	508.23	5987.92
5.	Murshidabad	16161.76	646.29	0	16808.05
6.	24-Parganas(N)	8641.45	1068.76	277.80	9988.01
7.	24-Parganas(S)	11237.79	1389.87	361.28	12988.94
8.	Howrah	4240.45	415.70	898.51	5554.66
9.	Hooghly	9224.22	4545.76	2498.03	16268.01
10	Medinipore	2886.55	11482.31	3830.27	37199.13
11.	Bankura	17553.56	3810.75	1332.70	22697.01
12.	Purulia	50078.06	14229.64	6947.37	71255.07
13.	Burdwan	20618.79	7386.63	3189.49	
14.	Birbhum	15260.62	596.57	413.54	17270.73
	State Total	194113.88	55162.95	26925.06	276201.89

Source: ME & MS Wing, Directorate of Fisheries, Government of West Bengal.

Annexure 8

District-wise Impounded Water Area (Ponds & Tanks) in West Bengal (in hectare)

Sl.No.	*District*	*River*	*Canal/Khal*	*Beel/Baor*	*Total area*
1.	Coochbehar	13095.82	1664.58	1168.11	15928.51
2.	Jalpaiguri	16006.44	562.29	504.38	17073.11
3.	Siliguri	3395.63	683.94	683.94	4763.51
4.	Uttar Dinajpur/ Dakshin Dinajpur	4032.08	1774.30	3548.59	9354.97
5.	Malda	6497.6	2511.42	4551.55	13560.57
6.	Murshidabad	36283.23	10013.88	3790.69	50087.8
7.	Nadia	4191.28	2771.75	4271.61	11234.64
8.	24-Parganas(N)	14299.00	8712.00	8861.19	31872.19
9.	24-Parganas(S)	17704.98	13492.17	5749.47	36946.92
10.	Howrah	107.36	2019.82	118.28	2245.46
11.	Hooghly	4358.74	3714.94	3884.76	11958.44
12.	Purba Medinipur/ Paschim Medinipur	12774.12	8766.07	404.01	21944.2
13.	Bankura	15960.15	11711.04	1973.00	29644.19
14.	Purulia	3707.29	0	0	3707.29
15.	Burdwan	11316.64	5991.36	1939.91	19247.91
16.	Birbhum	795.63	5695.85	632.16	7123.64
	Total in West Bengal	164525.99	80085.71	42081.65	286693.35

Source: ME & MS Wing, Directorate of Fisheries, Government of West Bengal.

Annexure 9

District-wise area of (a) River, (c) Canal/ Khal, (c) Beel & Baor in West Bengal

Sl. No.	*District*	*Inland*		*Marine*	
		Fish	*Prawn*	*Fish*	*Prawn*
1.	Coochbehar	14033.32	—	—	—
2.	Jalpaiguri	7090.57	—	—	—
3.	Siliguri	2180.10	—	—	—
4.	Uttar Dinajpur	12526.78	—	—	—
5.	Dakshin Dinajpur	30670.26	—	—	—
6.	Malda	31654.29	50.00	—	—
7.	Murshidabad	56497.09	100.00	—	—
8.	Nadia	53041.80	120.00	—	—
9.	North 24-Parganas	101174.69	12300.00	—	—
10.	South 24-Parganas	130002.80	6700.00	66400.00	7540.00
11.	Kolkata	21500.00	—	—	—
12.	Howrah	19597.98	50.00	—	—
13.	Hooghly	58088.39	200.00	—	—
14.	Purba Medinipur	872010.00	7500.00	96260.00	11300.00
15.	Paschim Medinipur	40280.05	340.00	—	—
16.	Bankura	53560.05	—	—	—
17.	Purulia	71231.20	—	—	—
18.	Burdwan	81121.47	300.00	—	—
19.	Birbhum	44578.98	—	—	—
	Total	910840.00	27660.00	162660.00	18840.00

Source: ME & MS Wing, Directorate of Fisheries, Government of West Bengal.

Annexure 10

District-wise Production of Fish during 2001 – 2002 (in MT)

Sl. No.	*District*	*Inland*		*Marine*	
		Fish	*Prawn*	*Fish*	*Prawn*
1.	Coochbehar	15134.57	—	—	—
2.	Jalpaiguri	7992.61	—	—	—
3.	Siliguri	2299.45	—	—	—
4.	Uttar Dinajpur	26303.43	—	—	—
5.	Dakshin Dinajpur	19853.38	—	—	—
6.	Malda	31309.03	—	—	—
7.	Murshidabad	55169.86	—	—	—
8.	Nadia	52557.91	—	—	—
9.	24-Parganas(N)	78170.38	10238.35	—	—
10.	24-Parganas(S)	105693.69	5701.25	69554.96	7035.00
11.	Kolkata	21819.00	—	—	—
12.	Howrah	19137.45	—	—	—
13.	Hooghly	57757.90	—	—	—
14.	Medinipur	132681.18	6845.40	97000.04	10710.00
15.	Bankura	49744.16	—	—	—
16.	Purulia	78511.64	—	—	—
17.	Burdwan	88691.65	—	—	—
18.	Birbhum	50197.26	—	—	—
	Total	893024.55	22785.00	166555.00	17745.00

Source: ME & MS Wing, Directorate of Fisheries, Government of West Bengal.

Annexure 11

District-wise Production of Fish during 2000-2001 (in MT)

Sl. No.	*District*	*Inland*		*Marine*	
		Fish	*Prawn*	*Fish*	*Prawn*
1.	Coochbehar	14547.50			
2.	Jalpaiguri	7642.82			
3.	Siliguri	2199.34			
4.	Uttar Dinajpur	25347.32			
5.	Dakshin Dinajpur	19075.32			
6.	Malda	30152.63			
7.	Murshidabad	54549.86			
8.	Nadia	35200.00			
9.	24-Parganas(N)	7278.06	10208.35		
10.	24-Parganas(S)	104762.58	5671.25	68739.96	6934.00
11.	Kolkata	21800.00			
12.	Howrah	18976.42			
13.	Hooghly	57175.90			
14.	Medinipore	131756.64	6805.40	94925.04	10401.00
15.	Bankura	49051.61			
16.	Purulia	77802.33			
17.	Burdwan	79255.54			
18.	Birbhum	49861.81			
	Total	856435.68	22685.00	163665.00	17335.00

Source : ME & Ms Wing, Directorate of Fisheries, Government of West Bengal.

Annexure 12

Size-group Distribution of Vested Impounded Water Units and Area (Acres) of West Bengal

Districts	Size Classes											
	0.05-1.99 acres		2.00-2.50 acres		2.51-5.00 acres		5.01-19.99 acres		20.00 & above		Total	
	No	Area	No	Area	No	Area	No	Area	No	Area	No	Area
1	2	3	4	5	6	7	8	9	10	11	12	13
1. Cooch-behar	264	165.525	21	45.70	73	271.49	103	1009.25	57	6257.66	518	7749.62
	50.97%	2.14%	4.05%	0.59%	14.10%	3.50%	19.88%	13.02%	11.00%	80.75%	1.69%	18.78%
2. Jalpaiguri	80	46.95	6	13.00	31	119.62	31	320.98	11	453.20	159	953.75
	50.31%	4.92%	3.77%	1.36%	19.50%	12.55%	19.50%	33.65%	6.92%	47.52%	0.52%	2.31%
3. Darjeeling	—	—	—	—	—	—	—	—	—	—	—	—
4. West Dinajpur	4406	1594.31	187	413.97	238	811.04	73	616.18	27	1484.25	4931	4919.76
	89.35%	32.41%	3.79%	8.41%	4.83%	16.49%	1.48%	12.52%	0.55%	30.17%	16.11%	11.92%
5. Malda	10080	4344.12	126	271.67	190	649.94	198	1913.35	62	3173.33	10656	10352.41
	94.59%	41.96%	1.18%	2.63%	1.79%	6.28%	1.85%	18.48%	0.58%	30.65%	34.81%	25.09%
6. Murshi-dabad	—	—	—	—	—	—	—	—	—	—	—	—
7. Birbhum	604	325.60	23	52.14	32	131.11	8	63.53	—	—	667	572.38
	90.55%	56.89%	3.45%	9.11%	4.80%	22.90%	1.20%	11.10%	—	—	2.18%	1.39%
8. Nadia	333	325.60	12	26.57	22	80.46	38	303.40	55	2793.26	460	3529.29
	72.39%	9.23%	2.75%	0.75%	4.78%	2.28%	8.26%	8.60%	11.96%	79.13%	1.50%	8.55%

Contd.

Contd.

Districts	*Size*						*Classes*					
	0.05-1.99 acres		*2.00-2.50 acres*		*2.51-5.00 acres*		*5.01-19.99 acres*		*20.00 & above*		*Total*	
	No	*Area*	*No*	*Area*	*No*	*Area*	*No*	*Area*	*No*	*Area*	*No*	*Area*
1	2	3	4	5	6	7	8	9	10	11	12	13
9. Burdwan	70002	44501.79	4173	7001.28	3915	13288.21	730	5707.88	43	1972.38	78863	72471.51
	88.85%	61.41%	5.25%	9.66%	4.93%	18.34%	0.92%	7.87%	0.05%	2.72%	6.96%	13.28%
10. Howrah	68127	13189.63	45	101.74	58	213.22	—	—	—	—	68230	13504.59
	99.85%	97.67%	07%	0.75%	0.08%	1.58%	—	—	—	—	6.02%	2.48%
11. Midnapur	4835	2282.65	94	200.58	109	367.10	75	625.17	3	70.19	5116	3545.69
	94.51%	64.37%	1.84%	5.66%	2.13%	10.35%	1.46%	17.63%	0.05%	99%	16.71%	8.59%
											100.00%	100.00%
12. Bankura	858	593.01	13	30.77	33	131.53	39	300.38	5	306.64	948	1362.33
	90.51%	43.53%	1.37%	2.26%	3.48%	9.11%	4.11%	22.05	0.53%	22.51%	3.10%	3.30%
											100%	100%
13. Purulia	—	—	—	—	—	—	—	—	—	—	—	—
14. Hooghly	2203	1009.77	93	90.81	92	51.98	39	310.17	2	92.75	2375	1755.48
	92.76%	57.52%	1.64%	5.17%	3.87%	14.35%	1.64%	17.67%	0.09%	5.29%	7.76%	4.25%
											100. %	100. %
15. North 24	1005	291.03	28	63.55	69	247.45	39	373.72	34	2646.99	1175	3622.74
Parganas	85.53%	8.03%	2.39%	1.75%	5.87%	6.83%	3.32%	10.32%	2.89%	73.07%	3.84%	8.78%
											100%	100%
16. South 24	733	188.31	2	4.38	5	16.34	2	12.92	—	—	742	2219.95
Parganas	98.78%	84.85%	0.27%	1.97%	0.68%	7.36%	0.27%	5.82%	—	—	2.42%	0.55%
											100%	100%
Total	—	—	—	—	—	—	—	—	—	—	30613	41264.72
											100.00%	100.00%

Annexure 13

Contribution of Fisheries Towards State Domestic Product (SDP) from 1994-95 to 2001-2002 in West Bengal

Year	*1994-95*	*1995-96*	*1996-97*	*1997-98*	*1998-99*	*1999-00*	*2000-01*	*2001-02*
West Bengal	56246.71	67135.55	74422.15	89594.62	106169.52	116898.62	128729.02	143537.68
Agriculture	17637.86	20453.93	23311.72	29124.98	33694.76	34339.98	34684.03	37583.06
Fishery	2035.77	2449.81	2778.02	3133.99	3496.22	3980.67	4628.70	4769.19
Forestry	487.63	612.65	724.66	821.92	1027.00	996.67	1055.99	1006.60

Source: Economic Review 2002-03, Government of West Bengal.

Annexure 14

District-wise Breakup of Tank Fisheries under Alternative Organizational Forms (figures are for 2003-04)

Name of the Districts	*Area under Culture Fishery (in Hectares)*								
	Tank Fishery								
	D.C. on composite fish culture	*C.D. Scheme*	*Minikit distribution*	*Inland Fisheries project*	*Other Private Tanks*	*Co-operative Sectors*	*Sewage-Fed Fisheries*	*Paddy-cum-Fisheries*	*Total*
1	2	3	4	5	6	7	8	9	10
1. Coochbehar	5 (0.11)	96 (2.09)	80 (1.74)	—	4.086 (89.08)	320 (6.98)	—	—	4587(100.00)
	(5.56)	(3.76)	(5.63)	—	(1.54)	(7.69)	—	—	(1.49)
2. Jalpaiguri and	5 (0.03)	120 (0.59)	62 (0.31)	—	20 000 (99.07)	—	—	—	20187 (100.00)
Darjeeling	(5.56)	(4.71)	(4.36)	—	(7.53)	—	—	—	(6.57)
3 W. Dinajpur	9 (0.06)	16 (0.11)	80 (0.57)	250 (1.79)	13609 (97.18)	40 (0.29)	—	—	14004 (100.00)
	(10.00)	(0.63)	(5.63)	(7.35)	(5.13)	(0.96)	—	—	(4.56)
4. Malda	5 (0.05)	14 (0.15)	80 (0.84)	150 (1.58)	9217 (97.17)	20 (0.21)	—	—	9486 (100.00)
	(5.56)	(0.55)	(5.63)	(4.41)	(3.47)	(0.48)	—	—	(3.09)
5. Murshidabad	7 (0.04)	240 (1.42)	80 (0.47)	300 (1.78)	15979 (94.63)	280·(1.66)	—	—	16886 (100.00)
	(5.56)	(0.55)	(5.63)	(4.41)	(6.02)	(6.73)	—	—	(5.50)
6. Birbhum									
7. Nadia	3 (0.04)	192 (2.69)	80 (1.12)	100 (1.40)	5541 (77.65)	1220 (17.1)	—	—	7136 (100.00)
	(3.33)	(7.53)	(5.63)	(2.94)	(2.09)	(29.33)	—	(67.90)	(5.50)
8. Burdwan									

Contd.

Contd.

Name of the Districts	*Area under Culture Fishery (in Hectares)*								
	Tank Fishery								
	D.C. on composite fish culture	*C.D. Scheme*	*Minikit distribution*	*Inland Fisheries project*	*Other Private Tanks*	*Co-operative Sectors*	*Sewage-Fed Fisheries*	*Paddy-cum-Fisheries*	*Total*
1	2	3	4	5	6	7	8	9	10
9. Howrah	7 (0.12) (7.78)	120 (2.14) (4.71)	80 (1.43) (5.63)	100 (1.78) (2.94)	5181 (92.39) (1.95)	120 (2.14) (2.89)	— —	— —	5608 (100.00) (1.82)
10. Midnapur									
11. Bankura									
12 Purulia									
13 Hooghly									
14 North 24 Parganas	5 (0.01) (5.55)	216 (0.63) (8.47)	100 (0.29) (7.03)	— —	9667 (28.20) (3.64)	1020 (2.97) (24.52)	— (19.56)	23.286 (89.13)	34294(100.00) (11.16)
15. South 24 Parganas	7 (0.03) (7.78)	168 (0.82) (6.59)	100 (0.49) (7.03)	— —	12438 (60.81) (4.68)	900 (4.40) (21.64)	4000 (100.00)	2840 (10.87)	20453(100.00) (6.66)

Annexure 15

Operational Inputs, Yield, Cost and Income from Composite Fish Culture Operations of High Level of Input Technology at Current Price Level

Item	*Quantity/ hectare*	*Cost /hectare*
Fingerlings	10000	4950
Mahua Oil Cake @ Rs. 2.50 /kg	2000	5000
Lime @ Rs. 1.50 /kg	500	750
Organic Fertilizer	6000	900
Chemical Fertilizer	500	1400
Feed	19000	47500
Weeds (collection charges @ Rs. 25/t)	120t	3000
Wages @ Rs. per month for 2 persons		7200
Depreciation of nets (3 years life)		3333
Pond maintenance		1000
Miscellaneous Operational Cost		2000
Interest on Operating Capital @ 18% half yearly		6820
Total		Rs. 84000

Total Cost = Fixed Cost + Operating Cost = Rs. 26,000 + Rs. 84,000 = Rs. 1,10,000
Average Gross Yield = 9,500 kg/ ha
Gross Income @ Rs. 20/kg = Rs. 1,90,000
Net Profit = Rs. 1,90,000 – Rs. 1,10,000 = Rs. 80,000

Source: M. Randhir, "Role of Economics in the Context of Aquaculture Development", Central Institute of Freshwater Aquaculture, 1994.

Annexure 16

Size-group Distribution of Non-vested Impounded Water Units in West Bengal

Districts	Size Classes											
	0.05-1.99 acres		*2.00-2.50 acres*		*2.51-5.00 acres*		*5.01-19.99 acres*		*20.00 & above*		*Total*	
	No	*Area*	*No*	*Area*	*No*	*Area*	*No*	*Area*	*No*	*Area*	*No*	*Area*
1	*2*	*3*	*4*	*5*	*6*	*7*	*8*	*9*	*10*	*11*	*12*	*13*
1. Cooch-behar	5854 96.59%	1390.28 47.69%	39 0.64%	84.79 2.90%	94 1.55%	330.96 11.35%	62 1.02%	560.83 19.23%	12 0.20%	549.03 18.83%	6061 (0.52%) 100.00%	2915.89 (0.54%) 100.00%
2. Jalpaiguri	5787 98.90%	1093.11 78.25%	8 0.15%	17.00 1.22%	40 0.68%	135.35 9.69%	16 0.27%	151.52 10.84	— —	— —	5851 (0.51%) 100.00%	1396.98 (0.26%) 100.00%
3. Darjeeling	NA	NA	NA	NA	NA	NA	NA	NA	NA	NA	198 (0.17%)	404.42 (0.07%)
4. West Dinajpur	31740 97.22%	23999.22 80.02%	608 1.14%	1344.42 4.49%	618 1.17%	1086.23 6.95%	186 0.36%	1544.07 5.15%	25 0.05%	1015.73 3.39%	53177 (4.70%) 100.00%	29989.67 (5.50%) 100.00%
5. Malda	18156 96.05%	8098.48 63.77%	247 1.31%	551.20 4.34%	312 1.65%	1077.80 8.49	151 0.80%	1349.27 10.62%	37 0.19%	1621.92 12.77%	18903 (1.66%) 100.00%	12698.66 (2.33%) 100.00%

Contd.

Contd.

Districts	*Size Classes*											
	0.05-1.99 acres		*2.00-2.50 acres*		*2.51-5.00 acres*		*5.01-19.99 acres*		*20.00 & above*		*Total*	
	No	*Area*	*No*	*Area*	*No*	*Area*	*No*	*Area*	*No*	*Area*	*No*	*Area*
1	*2*	*3*	*4*	*5*	*6*	*7*	*8*	*9*	*10*	*11*	*12*	*13*
6.Murshidabad	NA	NA	NA	NA	NA	NA	NA	—	NA	NA	70835	41715.90
											(6.25%)	(7.66%)
7. Birbhum	54136	33760.04	1112	2473.53	1354	4354.55	147	1057.21	1	32.50	56750	41677.83
	95.39%	81.01%	1.96%	5.96%	2.93%	10.44%	0.27%	2.54%	Neg	0.08%	(5.01%)	(7.64%)
											100.00%	100.00%
8. Nadia	12882	5997.81	264	269.43	292	950.83	130	1376.52	52	2503.04	136.20	11097.63
	94.57%	54.04%	1.95%	2.43%	2.14%	8.57%	0.96%	12.40%	0.38%	22.55%	(1.20%)	(2.03%)
											100.00%	100.00%
9. Burdwan	70002	44501.79	4173	7001.28	3915	13288.21	730	5707.88	43	1972.38	78863	72471.54
	88.85%	61.41%	5.25%	9.66%	4.93%	18.34%	0.92%	7.87%	0.05%	2.72%	(6.96%)	(13.28%)
10. Howrah	68127	13189.63	45	101.74	58	213.22	—	—	—	—	68230	13504.59
	99.85%	97.67%	07%	0.75%	08	1.58%	—	—	—	—	(6.02%)	(2.48%)
											100.00%	100.00%
11. Midnapur	389016	80707.00	873	1938.14	1088	3660.77	256	2030.25	—	—	391233	88336.16
	99.43%	91.36%	0.22%	2.20%	028%	4.14%	0.07%	2.30%	—	—	391233	88336.16
											100.00%	100.00%
12. Bankura	42028	25837.86	2270	4966.14	3321	983.82	1117	9208.09	29	1350.60	4865	42346.51
	86.20%	61.01%	4.65%	11.73%	6.81%	2.32%	2.29%	21.74%	50%	3.20%	(4.30%)	(7.76%)
											100.00%	100.00%
13. Purulia	NA	NA	NA	NA	NA	NA	NA	NA	NA	NA	22074	95609.72
											(1.94%)	(17.52%)

Contd.

Contd.

Districts	*Size*						*Classes*					
	0.05-1.99 acres		*2.00-2.50 acres*		*2.51-5.00 acres*		*5.01-19.99 acres*		*20.00 & above*		*Total*	
	No	*Area*	*No*	*Area*	*No*	*Area*	*No*	*Area*	*No*	*Area*	*No*	*Area*
1	*2*	*3*	*4*	*5*	*6*	*7*	*8*	*9*	*10*	*11*	*12*	*13*
14. Hooghly	77201 96.07%	27741.56 72.46%	1331 1.66%	2969.92 7.76%	1512 1.88%	5106.57 13.34%	303 0.39%	2465.74 6.44%	— —	— —	80347 (7.09%) 100.00%	382837.79 (7.02%) 100.00%
15. North 24 Parganas	70859 99.18%	15806.13 73.74%	196 0.27%	418.14 1.95%^	235 0.32%	930.14 4.34%	119 0.16%	1183.98 5.52%	56 0.07%	3096.74 14.45%	71465 (6.31%) 100.00%	21435.13 (3.93%) 100.00%
16. South 24 Parganas	143876 99.43%	26474.17 83.91%	287 0.20%	569 1.80%	379 0.27%	1336.97 4.23%	87 0.06%	780.64 2.46%	52 0.04%	2400.60 7.60%	144861 (12.78%) 100.00%	31561.77 (5.79%) 100.00%
Total											1131053 (100.00%)	545446.19% (100.00%)

Annexure 17

Allocation and Utilization of Fund in State Plan Budget (West Bengal)

	Plan period	*Year*	*Sanction*	*Expenditure*	*Percentage*
1.	First Five Year Plan	1951-56	154.47	40.50	26.32
2.	Second Five Year Plan	1956-61	75.74	60.32	79.24
3.	Third Five Year Plan	1961-66	204.55	111.64	54.50
4.	Annual Plan	1966-67	55.00	63.90	79.82
5.	Annual Plan	1967-68	55.00	41.86	75.30
6.	Annual Plan	1968-69	55.99	43.55	79.19
7.	Fourth Five Year Plan	1969-74	247.76	255.19	92.88
8.	Fifth Five Year Plan	1974-79	1049.79	654.16	79.46
9.	Annual Plan	1979-80	447.00	226.49	50.67
10.	Sixth Five Year Plan	1980-81	510.00	326.27	63.97
		1981-82	560.00	269.36	48.10
		1982-83	351.00	281.06	75.76
		1983-84	450.00	431.54	95.90
		1984-85	730.00	674.14	92.35
11.	Seventh Five Year Plan	1985-86	750.00	737.75	98.36
		1986-87	810.00	780.14	96.31
		1987-88	850.00	811.02	95.41
		1988-89	900.00	801.79	89.08
		1989-90	945.00	778.62	82.39
12.	Annual Plan	1990-91	1100.00	939.25	85.39
13.	Annual Plan	1991-92	1155.00	652.97	56.53
14.	Eighth Five Year Plan	1992-93	206.00	364.71	177.04
		1993-94	800.00	629.94	78.74
		1994-95	1453.00	1391.57	95.00

References

Agro-Economic Research Centre (1986): *Use of Tanks For Minor Irrigation and Pisciculture in West Bengal*, Visva-Bharati.

Ali, M.H. (1990) : *Fishermen's Co-operations in Bangladesh,* Economic Affairs (Vol. 35, No. 4), pp. 277-282.

Ayyappan, S. *et al.* (2004): Fisheries Sector in India: Dimensions of Development, *Indian Journal of Agricultural Economics,* Vol. 59, No. 3, July–Sept., 2004, pp. 394-410.

Centre for Monitoring Indian Economy (2003): *Marine Products, Industry, Market Size and Shares*, August 2003.

Chakrabarty, S.K. (1968): Some Problems of Fish Culture in West Bengal — a case study, *Indian Journal of Agricultural Economics,* Vol. 28, No. 1-4.

Chattopadhyay, K.S. (1993): *Factors Relating to Underdevelopment of Pisciculture in Birbhum District.* Occasional paper.

Chislam, M. (1962) : *Field demonstration of composite fish culture in West Bengal,* Fish Farmers' Development Agency Officials training in composite fish culture and fish seed production, CIFRI, Barrackpore.

Dehadri, P.V. (1994): Export Potential for Fishery Products, Ajit Singh (ed.). *Export of Agricultural Commodities,* New Delhi.

Dhondyal, S.P. and Singh, G.N. (1968): Benefit Cost Ratios in Fish Culture, *Indian Journal of Agricultural Economics,* Vol. XVIII.

F.F.D.A. Booklets: (2003-04 – 2002-03) (Birbhum, Mursidabad, Malda, Bankura Districts).

Ghorai, A.K. (1986): *Modernization of Small Scale Fish Seed Units, Quality Control and Marketing,* Second National Fish Seed Congress, Calcutta.

Government of India (2002): *Tenth Five-Year Plan*, 2002-2007, Vol. II, Planning Commission.

Heady'O. Earl and J.L. Dillon (1961): *Agricultural Production Function*, Iowa State University, Princeton.

Indian Journal of Agricultural Economics (1968): All articles in development of fisheries – Economic Aspects. Vol. 28, Jan.–March, No. 1.

Jana, R.K. and R.K. Sena (2004): Overwhelming Growth in Fisheries, The Hindu Survey of Indian Agriculture, p. 102.

Jose Murieker, S.C. (1983): Inter-linkage of credit and marketing relations in Fish Trade in U.K. Srivastava and M. Dharma Reddy (ed.) *Fisheries Development in India,* 1983, Concept Publishing Conpany, New Delhi.

Kamal, M.Yusuf (1994): Aquaculture Technology Development in India, Ajit Singh (ed.), *Export of Agricultural Commodities,* New Delhi.

Katiha, P.K. *et al.* (1997): Input Resource Management in Reservoir Fisheries, *Indian Journal of Agricultural Economics,* Vol. 52, No. 1 Jan.--March 1997, pp. 125-38.

Kumar, Anjani *et al.* (2004): Technical efficiency in Shrimp Farming in India: Estimation and Implication, *Indian Journal of Agricultural Economics,* Vol. 59, No. 3, July–Sept., 2004 pp. 413-20.

Kumar, H.D. (2003): *Sustainability and Management of Aquaculture and Fisheries*, Daya Publishing House, New Delhi.

Kumar, P. *et al.* (2004): A Study of Modelling of Household Demand for Fisheries in India, *Indian Journal of Agricultural Economics,* Vol. 59, No. 3, July–Sept., 2004.

Lawrence, R. Klein (1965): *An Introduction to Econometrics*, Princeton-Hall (Eastern Economy) Press, New Delhi.

Majumdar, T.C. (1985): *Fisheries in India.,* Akash Publication, New Delhi,

Marothia, D.K. (2002): *Alternative Governance in Common Pool Resources,* in *Institutionalizing Common Property Resources,* Concept Publishing Company, New Delhi.

Marothia, D.K. (2002): *Institutionalizing Common Property Resources*, Concept Publishing Company, Delhi.

Maurya, R.K. and Ravikesh (2000): Economics of Inland Fish Production – A Case Study of District Mau, UP. In *Indian Agricultural Marketing* (eds. J. Prasad and A. Prasad), Mittal Publications, New Delhi, India. pp. 271-77.

Misra, J. (2004): *Problems and Prospects of Inland Aquaculture in West Bengal – A Case Study*, Unpublished M.Phil. dissertation submitted to the University of Burdwan, West Bengal.

Misra, S.R. (1987): *Fisheries in India*. Ashis Publishing House, New Delhi.

Misra, S.R. and J. Misra (2005): "An Enquiry into the Nature and Constraints in the Development of Inland Fish Cultivation", Major Research Project, Sponsored by University Grants Commission, New Delhi. (Report Unpublished).

Mollah, A.R. *et al.* (1990): Input-output Relations in Fish Production under various Pond size, Ownership Pattern and Constraints, *Bangladesh Journal of Fisheries Development,* Vol. 3, No. 2.

Natarajan, A.V. (1985): Scope for Scientific and Commercial Fish Farming in West Bengal, *CIFRI Bulletin* No. 25.

National Council of Applied Economic Review (1992): *Export and Prospects of Fish and Fish Products*, New Delhi.

National Income Statistics (2002): Central Statistical Organization, Government of India.

Padhy, M.K. (1993): *Constraints in Pisciculture in Bolpur – Sriniketan Block – A case study* (an unpublished M.Sc. Thesis).

Pandey, A.C. and J.P. Misra (2000): Economic Feasibility of Fish Culutre in the District Faizabad (U.P.)—A Case Study. In *Indian Agricultural Marketing* (ed.) J. Prasad and A. Prasad, Mittal Publications, New Delhi, India, pp. 263-269.

Pandey A.C. (1993): Fish Farming – Importance and Governmental Assistance, *Unnat Krishi* 32.

Radheshyam, G. (2001): Commodity Based Agriculture in India – SWOT, *Aquabite,* ICLARM Quarterly, Vol. 24, Nov. 02, Jan.–June, 2001, pp. 9-15.

Rahim, K.M.B. and Singh, Katar (1992): *Central Fishermen's Co-operative Society, Contai—A case study.* A Project Report. Institute of Agriculture, Visva-Bharati and Institute of Rural Management, Anand, Gujarat.

Rahim, K.M.B. and Padhy, M.K. (1994): Scope and Constraints of Inland Pisciculture in West Bengal : A Case Study in Birbhum district, in the book *Role of Fisheries in Rural Development,* Edited by Giriappa S., Daya Publishing House, Delhi.

Ratna Reddy, V. *et al.* (2004): Ecological and Economic Aspects of Shrimp Farming in Andhra Pradesh, *Indian Journal of Agricultural Economics,* Vol. 59, No. 3, July–Sept., 2004.

Saxena, B.S. (1968): *Inland Fisheries in the National Economy.*

Sinha, M. *et al.* (2002): Management of Inland Fisheries Resources under Different Property Regimes in *Institutionalizing Common Property Resources,* Ed. D.K. Marothia, Concept Publishing Company, Delhi.

Singh, Ajit (Ed.)(1994): *Export of Agricultural Commodities 2000 A.D. – a perspective.* Wiley Eastern Limited, New Delhi.

Singh, B.B., Singh, R.K. and Yadav, R.M. (2000): Production and Marketing of Inland Fish: A Case Study in *Indian Agricultural Marketing* (eds.) J. Prasad and A. Prasad, Mittal Publications, New Delhi, India, pp. 279-87.

Singh, V.D. and Sampath, V. (1983): *Present Status, Policy Issues on Inland Fisheries.*

Shyam, S.S. *et al.* (2004): Export Performance of Indian Fisheries in the Context of Globalization, *Indian Journal of Agricultural Economics,* Vol. 59, No. 3, July–Sept., 2004.

Srivastava, U.K., Dholakia, B.H., Rao, S.S. and S. Vatsala (1990): *Evaluation Study of Fish Farmers' Development Agency Programme for Fresh Water Aquaculture*, Indian Institute of Management, Ahmedabad.

Srivastava, V.S. and S. Vatsala (1983): *Strategy for Development of Inland Fishery Resources in India – by Issues in Production and Marketing.*

World Bank Report (2002): *Socio-cultural Aspects of Developing Small Scale Fisheries*, Vol. 73.

World Bank Staff (2002): No. 490, 2002. Working paper.

Web Pages

Kurien John (2002): Small scale Fisheries in the Context of Globalization, CDS Working Papers, Centre for Development Studies, http://www.cds.ac.in.

Bene Cristopher and Neiland A.E. (2001): Contribution of Inland Fishery in the context of Rural Livelihood of Africa in http://www.lars2.org/unedited-papers

Payne, Ian (June 2000): The Changing Role of Fisheries in Development Policy, Natural Resource Perspectives (Nov. 59), http://www.odi.org.uk.

Srerdrup Sten (2000): Co-management in Small Scale Fisheries—A Synthesis of Southern & West African Experience, http://www.gt2.dc/fisheries-policies.

F.A.O. (2000): Small Scale Fisheries Development: Topic Fact Sheet. http://www.fao.org.

Bene, C., Minjdimka, K., Belal, E. and Jollet, T. (2001): Evaluating Livelihood Strategy and Role of Inland Fishery in Rural Development and Poverty Alleviation, http://www.oregonstate.edu/dept.

INDEX